**Industrial Control
Wiring Guide**

Industrial Control Wiring Guide
Second edition

Bob Mercer

Routledge
Taylor & Francis Group

LONDON AND NEW YORK

First published 2001 by Newnes

Published 2016 by Routledge
2 Park Square, Milton Park, Abingdon, Oxon, OX14 4RN
711 Third Avenue, New York, NY10017

Routledge is an imprint of the Taylor & Francis Group, an informa business

First edition 1995
Reprinted 1996, 1998, 1999
Second edition 2001
Reprinted 2002, 2004, 2005, 2006

British Library Cataloguing in Publication Data
A catalogue record for this book is available from the British Library

Library of Congress **Cataloging-in-Publication Data**
A catalog record for this book is available from the Library of Congress

ISBN-13: 978-0-7506-3140-2 (pbk)
ISBN-13: 978-1-1384-1340-5 (hbk)
ISBN-13: 978-0-8050-0861-0 (ebk)

CONTENTS

BS EN60529 – IP Codes

First number Protection against solid objects	**Second number** Protection against liquids
0. No protection	0. No protection
1. Protected against solid objects over 50 mm, e.g. accidental touch by hands	1. Protected against vertically falling drops of water
2. Protected against solid objects over 12 mm, e.g. fingers	2. Protected against direct sprays of water up to 15° from the vertical
3. Protected against solid objects over 2.5 mm (tools + wires)	3. Protected against sprays 60° from the vertical
4. Protected against solid objects over 1 mm (tools + small wires)	4. Protected against water sprayed from all directions – limited ingress permitted
5. Protected against dust – limited ingress (no harmful deposit) permitted	5. Protected against low pressure jets of water from all directions
6. Totally protected against dust	6. Protected against strong jets of water, limited ingress permitted
	7. Protected against the effects of immersion between 15 cm and 1 m
	8. Protected against long periods of immersion under pressure

1. SAFETY

1.1. Personal safety

Concern for your own safety as well as the safety of others should always be on your mind. Most safety procedures are common sense but, because some hazards are not obvious, there are regulations born out of experience which are designed to make the workplace safer.

There are two aspects of safety which concern us in the assembly of electrical equipment and control panels.

The first concerns your own personal safety. In the words of the *Health and Safety Regulations*:

- the need to use safe working practices and safety equipment to avoid the risk of injury to yourself and to others in the course of your work.

While it is beyond the scope of this book to cover the detail of all the safety precautions and safe working practices which should be adopted, there are some general points which can be noted.

- Safety equipment, e.g. goggles, gloves, etc., should be provided and *must be used* where they are appropriate.

- The onus is on you to use the safety equipment provided by your company. Any damage to safety gear should be reported. Safe working practices are part of any job and you should always learn and adopt them as a natural way of working.

- Don't take shortcuts which compromise your safety, or that of anybody else.

- You should make yourself aware of the procedures used at your place of work to prevent accidents and to deal with common incidents.

- You should know how to isolate electric supplies and how to work safely on electrical circuits.

1.1.1. Accidents

- Know how to contact the correct person – *the designated first aider* – for help.

- Find out the location of the nearest first aid box.

- Know how to isolate electric supplies and how to release a person safely from contact with electricity.

1. SAFETY

1.1.2. Fire

Before commencing work on electric plant, you should know:

- Where is the nearest fire alarm activator, fire exit and fire extinguisher?

- Are the fire exits clear of equipment or rubbish?

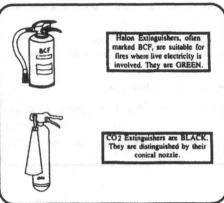

Halon Extinguishers, often marked BCF, are suitable for fires where live electricity is involved. They are GREEN.

CO2 Extinguishers are BLACK. They are distinguished by their conical nozzle.

1.1.3. Extinguishers for electrical fires

Be aware that *special extinguishers* are needed for fires which occur in *live electrical equipment* – do not use water-based extinguishers.

- RED extinguishers are water-based for wood/paper/cloth/plastic fires only.

- GREEN extinguishers are halon or BCF-based for general fires (not gases) including electrical fires.

- BLACK extinguishers are CO_2-based for flammable liquids and electrical fires.

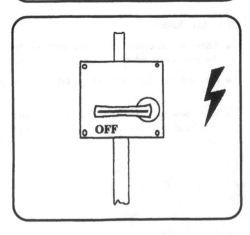

OFF

1.1.4. Electric shock

Learn the basic first aid action drill.

- *DO NOT TOUCH* the victim with your bare hands until the power is off or they have been pulled away from contact otherwise you will get a shock as well.

- Switch off the power and drag the victim off the live conductor.

- Alternatively if you cannot switch off then use something *non-conducting* to move the victim away from contact. Dry wood, plastic tubing (PVC conduit) even a dry piece of cloth folded several times will do.

1. SAFETY

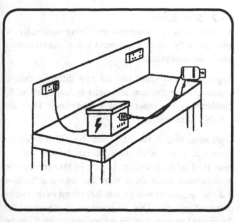

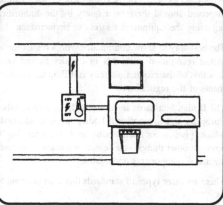

1.1.5. Working with electrical equipment

Many of the tools we use are electrically powered, some by the mains, some by battery.

- Mains-driven portable tools should be connected to the supply through an isolating transformer.

- These are usually 110 V systems which reduce the effect of electric shock.

- Heavier, fixed machines are wired into the three-phase factory supply. This is 415 V and there are stringent safety regulations governing its installation and use.

Remember . . .

- Do not take liberties or chances with electricity.

- Don't interfere with electrical apparatus, i.e. use it as intended and don't remove any covers or panels.

- Don't use or tamper with electrical machinery and tools that do not concern you. Leave switches and buttons alone. If you want to know how something works then ask someone who is authorised to show you.

- Take care when using portable electric power tools since these cause the highest number of accidents in the workplace. Accident possibilities range from tripping over a carelessly laid power cord to getting swarf in your eye because you didn't wear eye protection.

1. SAFETY

1.2. Building safe equipment

The second aspect of safety concerns the requirement to:

- design, construct and use electrical equipment so that it is safe and does not give rise to danger even should a fault occur.

The designer of the equipment will have taken into account all these concerns in specifying the parts to be used, the wire types and colours, the type of enclosure and so on. Our part comes in ensuring that:

- no parts are to be substituted without an engineering change notice;
- manufacturers' instructions for any component must be followed.

Both of these aspects are statutory requirements laid down in law in a number of regulations.

1.2.1. The Health and Safety at Work Act (HSAW)

This is a wide-ranging Act of Parliament covering all aspects of safety at work. It has gradually replaced the Factories Act.

The HSAW allows for the introduction of regulations to control particular aspects of safety at work. These regulations, which must be complied with, are often produced because of European Directives, which in turn are designed to harmonise the safe working conditions for all members of the European Community.

Among the many regulations within the Act, some have a direct influence on the machinery control panels which we are interested in, for example:

- The Electricity at Work Regulations 1989.
- The Provision and Use of Work Equipment.
- The Supply of Machinery (Safety) Regulations.
- The Electrical Equipment (Safety) Regulations.

These affect other areas of safety besides those which concern us here and it is outside the scope of this book to go into any real detail on them. However, you may find it useful to consider how they affect the way we build these panels and the components we use in them.

1.2.2. Standards

A standard is a document specifying nationally or internationally agreed properties for manufactured goods and equipment.

Regulations and standards are two different things: regulations are the law and must be complied with; standards on the other hand are advisory. They are closely linked together.

Equipment sold in the EEC must be 'CE marked' to show that it complies with the regulations that are concerned with its safety. As part of this process the manufacturer must show how the risks and hazards that the equipment will present have been overcome or protected against. This information is placed in the Technical Document of the equipment so that it can be inspected should there be a query by the authorities regarding the equipment's safety or performance.

The best way to show compliance with a regulation is to use recognised standards in the design and construction of the product, thereby fulfilling the requirements of the regulation.

The British Standards Institute (BSI), as well as other European and international bodies, publish standards which give recommendations and guidance on – amongst other things – the selection and use of various electrical components and cables.

There are three types of standards that are important to us:

- British Standards (BS),
- European Harmonised Standards (EN or BS EN),
- International Standards (IEC).

These are of course mainly the concern of the designer but it is as well to be aware that they exist, as it may explain why one component is used instead of another and why only those components designated in the parts list must be used.

Standards of most importance to us are:

- BS EN 60204 – Safety of Machinery – Electrical Equipment of Machines,
- BS EN 60947 (IEC947) – Low Voltage Switchgear and Controlgear (7 Parts).

1. SAFETY

BS EN 60204 covers the way in which the electrical equipment should be constructed and includes everything from the selection of components, through the sizes, types and colour of the wiring, to the electrical tests that should be done on the finished equipment. Within BS EN 60204 there are references to other standards, including BS EN 60947, that will give more detail on individual parts or components.

BS EN 60947 and the international standard IEC 947 are in seven parts, giving the specification and other requirements of the individual components we will use in the equipment.

- Part 1: General Requirements. Defines the rules of a general nature to obtain uniformity in requirements and tests.

Each of the following parts deals mainly with the characteristics, conditions for operation, methods for testing and marking requirements of the various electrical components.

- Part 2: Circuit-breakers.

- Part 3: Switches, disconnectors, switch-disconnectors and fuse combination units.

- Part 4: Contactors and motor starters including short circuit and overload protection devices.

- Part 5: Control circuit devices and switching elements.

- Part 6: Multiple function equipment such as that used for automatic emergency power switching.

- Part 7: Ancillary equipment such as terminal blocks used to connect copper conductors.

Basically our control equipment panels should be built to conform to the requirements of BS EN 60204 using components manufactured to conform to the requirements of BS EN 60947 and other related component standards and approvals.

An approved component is one whose manufacture and performance has been checked and proven to meet the specifications set by the standards authority of an individual country. For example, a part approved in the UK would be 'BS approved'. These approvals may be important if the equipment is to be exported.

Some other standards authorities are:

- USA ANSI – approvals are made by the Underwriters Laboratory and marked UL.

- Canada, CSA.

- Denmark, DEMKO.

- Italy, CEI.

- Norway, NEMKO.

- Germany, DIN/VDE.

- France, NF/UTE.

- Europe, CENELEC.

Some other BSI documents

- PD 2754: Parts 1 and 2. Published document. Construction of electrical equipment for protection against electric shock. Part 1 deals with the classification of electrical and electronic equipment with regard to protection against electric shock, for example whether it is earthed, double insulated or uses a safe, low voltage supply. Part 2 is a more detailed guide to the requirements of the various classes as defined in Part 1.

- BS 7452: Specification for transformers of the type used in control panels. Equivalent to IEC 989: Control transformer specification.

- BS 3939: Graphical symbols. Provides comprehensive details of the symbols to be used in electrical, electronic and telecommunication diagrams. It is published in 12 parts and is broadly the same as EN 617 – Parts 2 to 12.

- BS EN 60073: Colours for indicator lamps, push buttons, etc. Provides a general set of rules for the use of certain colours, shapes, positioning requirements of indicators and actuators to increase the safety and operational efficiency of equipment. BS EN 60204 also provides guidelines specific to the electrical controls for machinery.

- BS EN 60529: Specification for classification of degrees of protection provided by electrical enclosures. Also known as 'IP Codes', it uses a

1. SAFETY

two or three digit number to define to what degree the enclosure is sealed to protect the contents against dust, moisture and similar damaging substances.

● BS 6231: Specification for PVC-insulated cable for switchgear and controlgear wiring. This deals with the requirements for the wires and cables used in the wiring of control panels up to 600 V/1000 V.

1.2.3. The IEE regulations (BS 7671)

The Institute of Electrical Engineers publishes its Regulations for Electrical Installations, which cover the design, selection and construction of electrical installations in buildings generally, and provide guidance for safety in the design and construction of electrical equipment. Although mainly concerned with electrical systems in buildings, the information is applicable to machine control panels because they will be connected to the building's electrical system. These are now published as BS 7671.

In addition there are:

● Guidance Notes from the Health and Safety Executive.

● Specifications and Regulations from the Department of Trade and Industry, BSI and the Institute of Electrical Engineers.

You may be interested to know that all the standards referred to here concern 'low voltage' equipment. Low voltages as defined by the IEE are those up to and including 1000 volts AC or 1500 volts DC.

2. DRAWINGS

2.1. Types of drawing

We use drawings to convey the information about a piece of equipment in a form which all those involved in its production, installation and service will understand. To make this possible, standard drawing conventions have been adopted by most companies.

- This book will emphasise the British Standard symbols as defined in BS 3939. Other symbols which may be in common use will also be shown.

The information we need to be able to assemble the equipment will be only one item in the set of drawings and schedules which make up the complete design.

2.1.1. Circuit diagram

This shows how the electrical components are connected together and uses:

- symbols to represent the components;
- lines to represent the functional conductors or wires which connect them together.

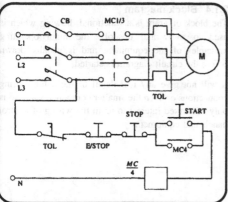

A circuit drawing is derived from a block or functional diagram (see 2.1.4.). It does not generally bear any relationship to the physical shape, size or layout of the parts and although you could wire up an assembly from the information given in it, they are usually intended to show the detail of how an electrical circuit works.

2.1.2. Wiring diagram

This is the drawing which shows all the wiring between the parts, such as:

- control or signal functions;
- power supplies and earth connections;
- termination of unused leads, contacts;
- interconnection via terminal posts, blocks, plugs, sockets, lead-throughs.

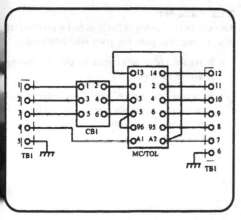

It will have details, such as the terminal identification numbers which enable us to wire the unit together. Parts of the wiring diagram may simply be shown as blocks with no indication as to the electrical components inside. These are usually sub-assemblies made separately, i.e. pre-assembled circuits or modules.

2. DRAWINGS

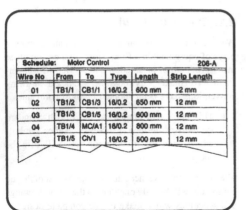

Schedule:	Motor Control				206-A
Wire No	From	To	Type	Length	Strip Length
01	TB1/1	CB1/1	16/0.2	600 mm	12 mm
02	TB1/2	CB1/3	16/0.2	650 mm	12 mm
03	TB1/3	CB1/5	16/0.2	600 mm	12 mm
04	TB1/4	MC/A1	16/0.2	800 mm	12 mm
05	TB1/5	Ch/1	16/0.2	500 mm	12 mm

2.1.3. Wiring schedule

This defines the wire reference number, type (size and number of conductors), length and the amount of insulation stripping required for soldering.

In complex equipment you may also find a table of interconnections which will give the starting and finishing reference points of each connection as well as other important information such as wire colour, ident marking and so on.

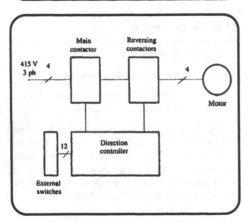

2.1.4. Block diagram

The block diagram is a functional drawing which is used to show and describe the main operating principles of the equipment and is usually drawn before the circuit diagram is started.

It will not give any real detail of the actual wiring connections or even the smaller components and so is only of limited interest to us in the wiring of control panels and equipment.

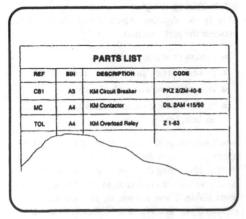

PARTS LIST			
REF	BIN	DESCRIPTION	CODE
CB1	A3	KM Circuit Breaker	PKZ 2/ZM-40-8
MC	A4	KM Contactor	DIL 2AM 415/50
TOL	A4	KM Overload Relay	Z 1-63

2.1.5. Parts list

Although not a drawing in itself, in fact it may be part of a drawing. The parts list gives vital information:

- It relates component types to circuit drawing reference numbers.

- It is used to locate and cross refer actual component code numbers to ensure you have the correct parts to commence a wiring job.

2. DRAWINGS

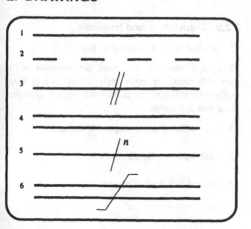

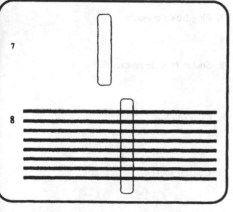

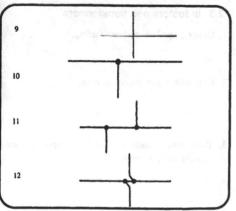

2.2. Symbols

2.2.1. Conductors

1. General symbol, conductor or group of conductors.

2. Temporary connection or jumper.

3. Two conductors, single-line representation.

4. Two conductors, multi-line representation.

5. Single-line representation of n conductors.

6. Twisted conductors. (Twisted pair in this example.)

7. General symbol denoting a cable.

8. Example: eight conductor (four pair) cable.

9. Crossing conductors – no connection.

10. Junction of conductors (connected).

11. Double junction of conductors.

12. Alternatively used double junction.

9

2. DRAWINGS

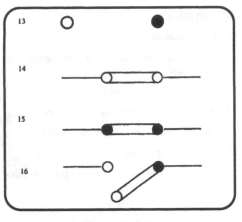

2.2.2. Connectors and terminals

13. General symbol, terminal or tag.

These symbols are also used for contacts with moveable links. The open circle is used to represent easily separable contacts and a solid circle is used for those that are bolted.

14. Link with two easily separable contacts.

15. Link with two bolted contacts.

16. Hinged link, normally open.

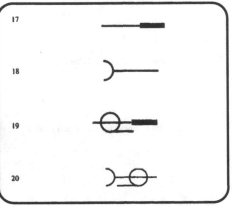

17. Plug (male contact).

18. Socket (female contact).

19. Coaxial plug.

20. Coaxial socket.

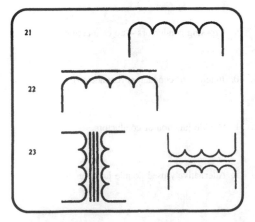

2.2.3. Inductors and transformers

21. General symbol, coil or winding.

22. Coil with a ferromagnetic core.

23. Transformer symbols. (See the components section for further variations.)

2. DRAWINGS

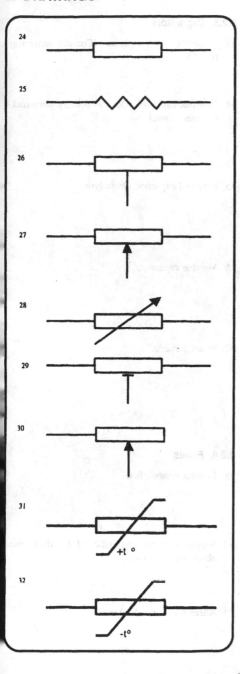

2.2.4. Resistors

24. General symbol.

25. Old symbol sometimes used.

26. Fixed resistor with a fixed tapping.

27. General symbol, variable resistance (potentiometer).

28. Alternative (old).

29. Variable resistor with preset adjustment.

30. Two terminal variable resistance (rheostat).

31. Resistor with positive temperature coefficient (PTC thermistor).

32. Resistor with negative temperature coefficient (NTC thermistor).

2. DRAWINGS

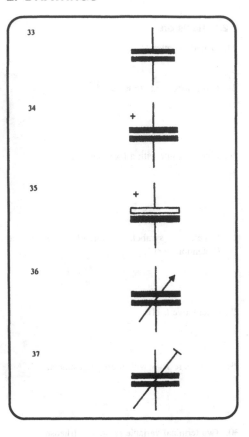

2.2.5. Capacitors

33. General symbol, capacitor. (Connect either way round.)

34. Polarised capacitor. (Observe polarity when making connection.)

35. Polarised capacitor, electrolytic.

36. Variable capacitor.

37. Preset variable.

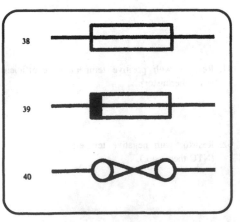

2.2.6. Fuses

38. General symbol, fuse.

39. Supply side may be indicated by thick line: observe orientation.

40. Alternative symbol (older).

2. DRAWINGS

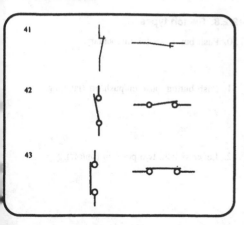

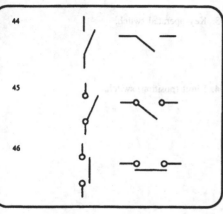

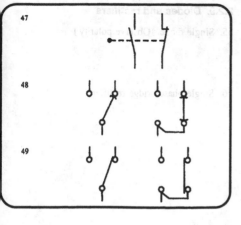

2.2.7. Switch contacts

41. Break contact (BSI).

42. Alternative break contact version 1 (older).

43. Alternative break contact version 2.

44. Make contact (BSI).

45. Alternative make contact version 1.

46. Alternative make contact version 2.

47. Changeover contacts (BSI).

48. Alternative showing make-before-break.

49. Alternative showing break-before-make.

13

2. DRAWINGS

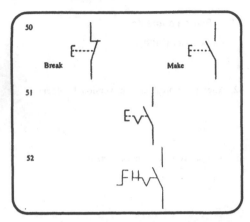

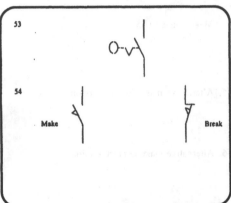

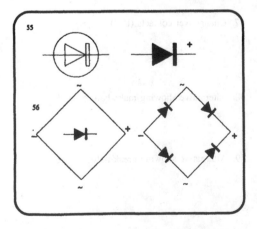

2.2.8. Switch types

50. Push button switch momentary.

51. Push button, push on/push off (latching).

52. Lever switch, two position (on/off).

53. Key-operated switch.

54. Limit (position) switch.

2.2.9. Diodes and rectifiers

55. Single diode. (Observe polarity.)

56. Single phase bridge rectifier.

2. DRAWINGS

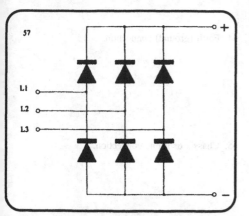

57. Three-phase bridge rectifier arrangement.

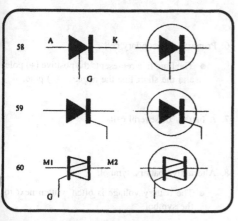

58. Thyristor or silicon controlled rectifier (SCR) – general symbol.

59. Thyristor – common usage.

60. Triac – a two-way thyristor.

2.2.10. Miscellaneous symbols

61. Direct Current (DC).

62. Alternating Current (AC).

63. Rectified but unsmoothed AC. Also called 'raw DC'.

15

2. DRAWINGS

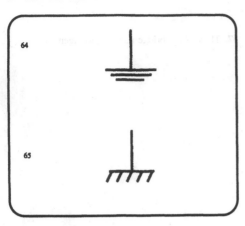

64. Earth (ground) connection.

65. Chassis or frame connection.

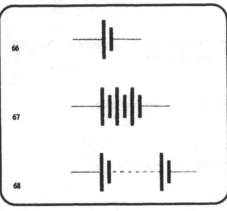

66. Primary or secondary cell.
  - The long line represents the positive (+) pole and the short line the negative (−) pole.

67. A battery of several cells.

68. Alternative battery symbol.
  - The battery voltage is often written next to the symbol.

3. WIRE TYPES AND PREPARATION

Introduction

Electrical equipment uses a wide variety of wire and cable types and it is up to us to be able to correctly identify and use the wires which have been specified. The wrong wire types will cause operational problems and could render the unit unsafe. Such factors include:

- the insulation material;
- the size of the conductor;
- what it's made of;
- whether it's solid or stranded and flexible.

These are all considerations which the designer has to take into account to suit the final application of the equipment.

A **conductor** is a material which will allow an electric current to flow easily. In the case of a wire connection, it needs to be a very good conductor. Good conductors include most metals. The most common conductor used in wire is copper, although you may come across others such as aluminium.

An **insulator** on the other hand is a material which does not allow an electric current to flow. Rubber and most plastics are insulators.

3.1. Insulation materials

Wires and cables (conductors) are insulated and protected by a variety of materials (insulators) each one having its own particular properties. The type of material used will be determined by the designer who will take into account the environment in which a control panel or installation is expected to operate as well as the application of individual wires within the panel.

As part of the insulating function, a material may have to withstand without failing:

- extremes of current or temperature;
- a corrosive or similarly harsh environment,
- higher voltages than the rest of the circuit.

Because of these different properties and applications, it is essential that you check the wiring specification for the correct type to use.

PVC (Polyvinylchloride)

This is the most commonly used general-purpose insulation. It will soften at higher temperatures and will permanently deform. Temperature range is −20°C to +75°C. This means that a soldering iron will melt it easily.

Polythene

A wax-like, translucent material which is used mainly for high voltage and high frequency applications.

PTFE (Polytetrafluoroethylene)

Similar to polythene but used for higher temperature environments (up to about +250°C).

Silicone rubber

Appears similar to natural rubber but feels smoother. It is used in harsh environments where elevated temperatures, radiation or chemical vapours are encountered.

Polyurethane

Generally found as a thin coating on copper wire. Used in transformer windings and similar applications. Some are 'self fluxing' during soldering but *may give off harmful fumes*.

Enamel

Used like polyurethane as a thin layer on copper wires.

Glass fibre

Usually woven it is used for extremely high temperature applications. *Wear gloves when using glass fibres; they are a skin irritant.*

Other types

There are other less common materials used in some specialised cables and you should become familiar with those used at your workplace. Some wires are insulated with Low Smoke and Fume (LSF) materials, the use of which is self-evident. These are halogen free, with Polyolefin and Polyethylene being two common materials.

3. WIRE TYPES AND PREPARATION

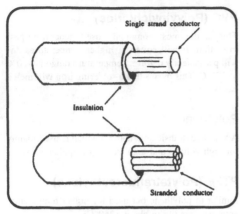

Single strand conductor

Insulation

Stranded conductor

3.2. Conductors

The conductor can be a single solid wire or made up of a number of thin strands.

- Solid or single-stranded wire is not very flexible and is used where rigid connections are acceptable or preferred – usually in high current applications in power switching contractors. It may be uninsulated.

- Stranded wire is flexible and most interconnections between components are made with it.

- Braided wire: see Sections 3.5 and 9.1.2.

3.3. Wire specifications

There are several ways to describe the wire type. The most used method is to specify the number of strands in the conductor, the diameter of the strands, the cross-sectional area of the conductor then the insulation type.

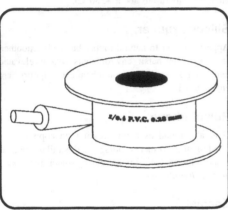

Example 1:

- The **1** means that it is single conductor wire.

- The conductor is **0.6 mm** in diameter and is insulated with **PVC**.

- The conductor has a cross-sectional area nominally of **0.28 mm²**.

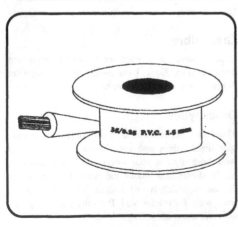

Example 2:

- The conductor comprises **35** strands.

- Each strand is **0.25 mm** and is insulated with **PVC**.

- The conductor has a total cross-sectional area nominally of **1.5 mm²**.

As well as this size designation the insulation colour will often be specified.

18

3. WIRE TYPES AND PREPARATION

SWG table

SWG No.	Diameter
14 swg	2 mm
16 swg	1.63 mm
18 swg	1.22 mm
20 swg	0.91 mm
22 swg	0.75 mm
24 swg	0.56 mm
25 swg	0.5 mm
30 swg	0.25 mm

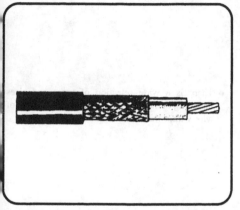

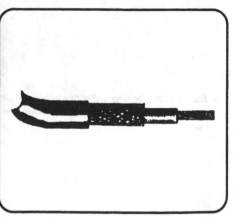

3.4. Standard Wire Gauge

Solid wire can also be specified using the Standard Wire Gauge or SWG system.

- The SWG number is equivalent to a specific diameter of conductor.
- For example; 30 SWG is 0.25 mm diameter.
- 14 SWG is 2 mm in diameter.
- The larger the number – the smaller the size of the conductor.

There is also an American Wire Gauge (AWG) which uses the same principle, but the numbers and sizes *do not* correspond to those of SWG.

3.5. Coaxial and screened wire

3.5.1. Coaxial
Coaxial cable has:

- an insulated central conductor surrounded by an outer tubular conductor;
- an outer conductor which is usually braided (woven) to give the cable flexibility;
- insulation between the two conductors which may be solid polythene, cellular polythene, polythene spacers, solid PTFE.

Although relatively expensive, it has low electrical losses and is used for the transmission of high frequency signal currents such as those found in high speed data transmission and radio systems. A common example is the cable between a television set and the aerial.

3.5.2. Screened
Screened wire is an *ordinary* insulated conductor surrounded by a conductive braiding. In this case the metal outer is not used to carry current but is normally connected to earth to provide an electrical shield to screen the internal conductors from outside electro-magnetic interference.

Screened wiring is generally only used for DC and lower frequency signals such as audio. It is often used for the input connections of PLCs where the voltage and current levels are low. These low level signals may need to be screened from the interference generated by cables carrying higher power voltages and currents.

3. WIRE TYPES AND PREPARATION

3.6. Multiway cables

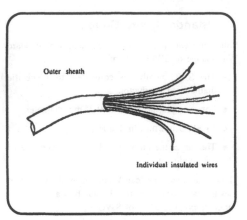

Outer sheath

Individual insulated wires

- Multiway or multicore cables have a number of individual insulated wires enclosed in an outer sheath.

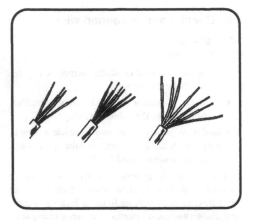

- There is a wide selection of types and sizes including some with a mix of different types of wire within the outer sheath.

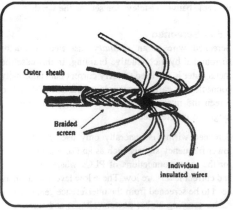

Outer sheath

Braided screen

Individual insulated wires

- The cable may be screened with a braiding made from tinned copper, steel wire or aluminium tape.

3. WIRE TYPES AND PREPARATION

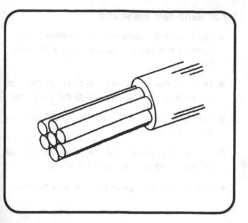

3.7. Insulation removal

Introduction

The removal of insulation from wires and cables is one of those tasks which, like soldering or crimping, is a major part of assembly work. There are many techniques used within the industry, using tools ranging from the simple hand-operated strippers to automatic, motorised types.

Hand-operated strippers fall into two main categories: those which are adjustable and those which are not. Within the non-adjustable types are some which have flexible jaws and will strip a range of wire sizes, while others have a series of cutting holes for each wire size.

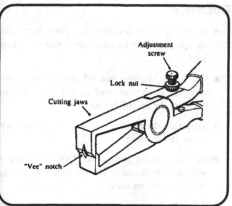

3.7.1. Adjustable hand tool

- These have jaws with V-shaped notches to cut the insulation.

- The adjuster screw acts as a stop to allow for a range of wire diameters.

- Adjust the screw to open or close the jaws so that the V cutting slots cut the insulation cleanly without tearing the insulation or damaging the conductor.

- Use a test piece of wire to adjust the jaws to the correct position to cut the insulation but not the conductor.

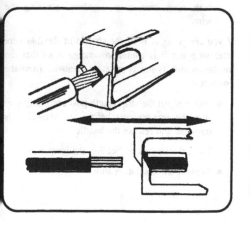

- Place the wire in the lower groove, squeeze the handles to cut the insulation, rotate the strippers half a turn and pull off the insulation stub.

- Check for damage to the conductor.

- When the adjustment is found to be correct, tighten the lock nut and test again. If OK, then the strippers are ready for use.

- *Always check* the wire for damage each time you remove insulation with this type of wire stripper.

3. WIRE TYPES AND PREPARATION

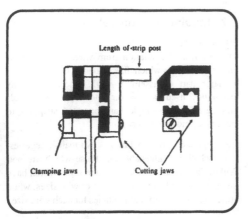

Length of strip post

Clamping jaws   Cutting jaws

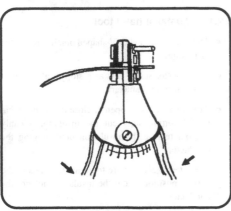

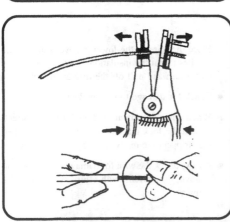

3.7.2. Hand-held automatic

- These are fully automatic in operation but it is essential that you use the correct size of cutting hole.

- There are two sets of jaws: one clamps the wire and holds it while the other cuts the insulation.

- Both jaws separate to pull the insulation stub away from the wire.

- The cutting blades can be changed to suit different sizes of conductor diameters.

- A 'length of strip' guide post can also be fitted.

Operation

- Place the wire between the jaws from the clamping jaw side into the correct size of cutting notch.

- If a 'length of strip' post is fitted the end of the wire should be positioned so that the end is in line with the end of the post.

- Squeezing the handles will first cause the wire clamp jaw to close.

- Next the cutting jaws close; further squeezing will cause both sets of jaws to separate, pulling off the insulation stub.

- Continue to squeeze the handles and the jaws both open then snap together, releasing the wire.

If you are going to twist the strands of flexible wire after stripping it is useful to arrange it so that the insulation stub is not completely removed from the conductor.

- Either adjust the strip length post accordingly or stop the process just before the insulation is removed and release the handles.

- Twist the strands by holding the insulation.

- Remove the insulation stub.

3. WIRE TYPES AND PREPARATION

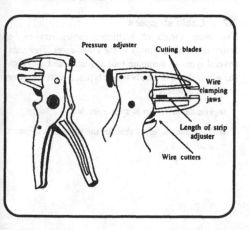

3.7.3. Non-adjustable

- These have no adjustment for the wire size, though there are adjustments for length of strip and jaw pressure.

- The jaws are designed to firmly grip the insulation without marking it.

- Adjust the strip length as required.

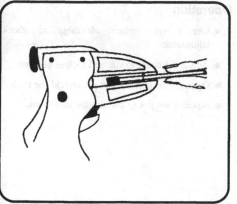

- Place the wire between the jaws so that it touches the strip length adjuster.

- Squeeze the handles and the jaws grip the wire.

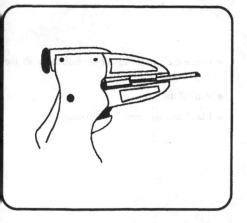

- Further pressure and the cutters move to pull the insulation off.

- If you are going to twist the wire, adjust the length so that the insulation stub is not removed. Twist the strands using the stub.

On some tools the cutting blades are flexible and form themselves around the conductor as they cut through the insulation, which is then pulled away by the action of the jaws.

3. WIRE TYPES AND PREPARATION

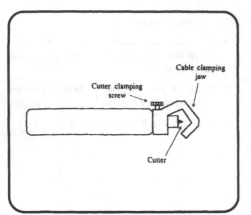

3.7.4. Cable strippers

The outer sheath of multicore cables has to be removed without damage to the inner cores. There are several types of stripping tool available and although the actual detail differs between types, a representative tool is shown here.

- The cable clamp is spring loaded.

- The cutter is adjustable for the thickness of outer insulation.

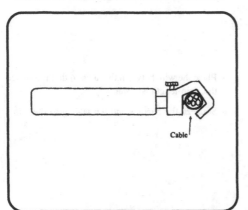

Operation

- Use a spare piece of cable to check adjustments.

- Adjust the cutter to suit and lock in position.

- Open up the jaw and place it around the cable.

- Squeeze the jaw to cut into the insulation.

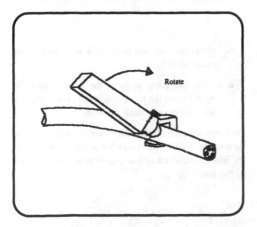

- Rotate the stripper to cut the insulation all the way round.

- Pull off the insulation.

- Check that no inner core is damaged.

3. WIRE TYPES AND PREPARATION

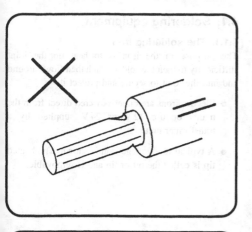

3.7.5. Fault prevention

Here are some examples of the damage that can be caused by lack of care when removing insulation.

- Nicks in solid conductors.

- Stripped plating and scores in solid metal conductors.

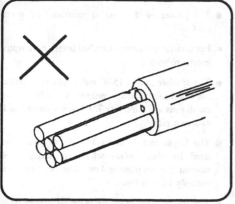

- Strands cut out of multistrand wires.

- Strands nicked.

These cause the wire to be mechanically weakened and its current carrying capacity reduced.

Each of these has the potential to cause the finished equipment to malfunction when it is in service, because the wire will eventually break off or even act like a fuse.

Summary

Hand strippers are a common cause of damage to insulation and conductors so you must frequently check the adjustment and, with the so-called automatic type, be sure to use the correct hole size.

It is all too easy to place the wire into the wrong hole, either smaller, causing the damage already mentioned, or larger, which can tear the insulation leaving a ragged edge which may get mixed in with the solder and cause contamination of the joint.

Most of these problems can go unnoticed except by you at the time they occur. However, remember that they will cause operational problems to the end-user, so don't let them pass.

Whatever type you are using, read the instruction leaflet which the manufacturer provides before using them. If you use an adjustable type, be sure to adjust it properly and check it regularly in use.

4. SOLDERING AND TERMINATION

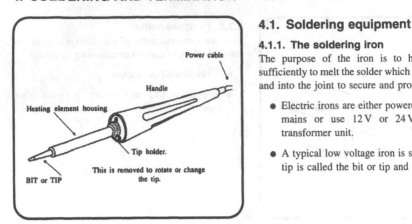

Power cable

Handle

Heating element housing

Tip holder.

BIT or TIP

This is removed to rotate or change the tip.

4.1. Soldering equipment

4.1.1. The soldering iron

The purpose of the iron is to heat up the joint sufficiently to melt the solder which then flows around and into the joint to secure and protect it.

- Electric irons are either powered direct from the mains or use 12 V or 24 V supplied by a transformer unit.

- A typical low voltage iron is shown. The heated tip is called the bit or tip and is removable.

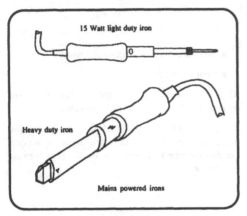

15 Watt light duty iron

Heavy duty iron

Mains powered irons

- The power of the iron is measured in watts (W).

- For non-temperature controlled irons, more watts means more heat.

- The smaller iron is 15 W and would be suitable for small joints such as printed circuit boards or small pins and wire. Say 7/0.2 wire soldered to 1 or 2 mm pins.

- The larger iron is about 100 W and would be used for those joints which are larger than normal. Say tin plate or 4 mm diameter wire to a suitably large solder tag.

- The *temperature of the tip* is the most important factor, so the normal iron is temperature controlled between 250°C and 400°C, and for safety is powered by a low voltage supply.

- The temperature control may be incorporated in the base unit and varied by a control knob.

- In other irons the tip itself determines the operating temperature. To change the temperature, you change the tip.

- An enclosing holder and a sponge may be incorporated into the base unit or as a separate unit as in this example.

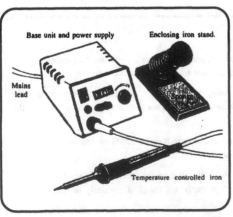

Base unit and power supply

Enclosing iron stand.

Mains lead

Temperature controlled iron

4. SOLDERING AND TERMINATION

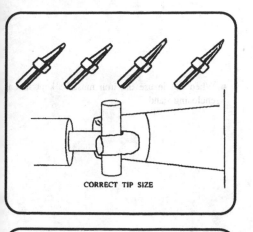

CORRECT TIP SIZE

- Tips or bits come in a variety of shapes. All modern tips are plated to prolong their operational life.

- The tip must be tinned before being used for the first time. This simply means melting a little solder on to it once it is hot enough.

- Use a tip size and shape which will allow the tinned end to *touch both parts of the joint*.

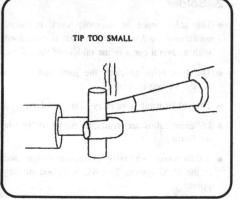

TIP TOO SMALL

- This one is too small and will not heat the joint enough to melt the solder.

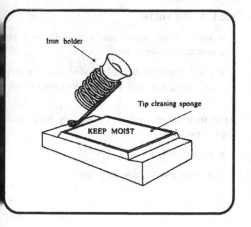

Iron holder

Tip cleaning sponge

KEEP MOIST

- The sponge is dampened and used to clean the tip.

- DO NOT use a wire brush or file on plated tips.

4. SOLDERING AND TERMINATION

- When not in use the iron must be kept in an enclosing stand.

4.2. Solder

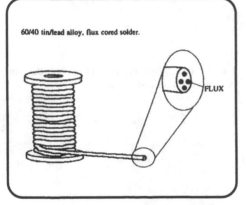

60/40 tin/lead alloy, flux cored solder.

FLUX

- The solder used in assembly work is called *multicored solder* since the flux is contained within several cores in the middle of the solder.

- The flux helps to clean the joint and should *always* be non-corrosive.

- The solder itself is an alloy of tin and lead.

- Different ratios are available. 60/40 tin/lead is the norm.

- Solder comes in a variety of diameters expressed in the SWG system. 20 SWG is a good starting point.

4.2.1. Using solder

- Most of the joints you will make will be connecting wire to pins.

- Practise melting the solder and making a solder joint using a piece of 22 SWG bare tinned copper wire.

Apart from the iron and solder, the only basic tools needed are:

- Wire cutters;

- Smooth jaw, snipe-nosed pliers.

4. SOLDERING AND TERMINATION

4.3. Forming the wire

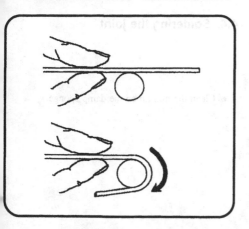

It is essential in this kind of connection to *make a good mechanical joint* before soldering. This then takes any strain rather than the solder having to do so. The solder's job is mainly to protect the joint from the atmosphere.

- The parts to be soldered must be *clean* and free from grease. Avoid touching them with your fingers.

- Place the wire against the pin.

- Use the pliers to form it round the pin.

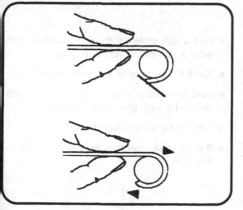

- Trim off the excess.

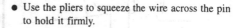

- Use the pliers to squeeze the wire across the pin to hold it firmly.

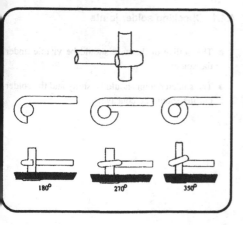

180°   270°   350°

- You should now have a joint which looks similar to this.

- The amount the wire is wrapped around the pin can be anywhere between 180° and 350° depending upon the application of the finished unit.

- Aerospace and defence work, for example, requires 350°.

29

4. SOLDERING AND TERMINATION

4.4. Soldering the joint

- Clean the iron tip on the damp sponge.

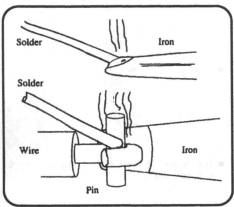

- Melt a little solder on the tip of the iron. This helps to transfer the heat to the joint.
- Touch both parts to be soldered – wire and pin.
- Feed the solder in from the opposite side. It will melt and quickly flow around the joint.
- Remove the solder *before* the iron.
- It should take about three seconds to heat, melt and flow.

4.4.1. Checking solder joints

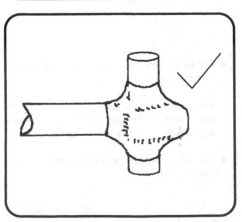

- The outline of the wire should be visible under the solder.
- The soldered joint should be shiny and the solder outline should be concave.

4. SOLDERING AND TERMINATION

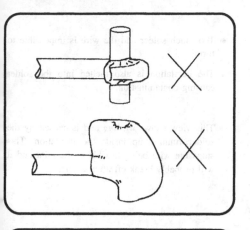

- This joint is not covered. There is not enough solder. Reheat and put more on.

- There is too much solder here. Use a desoldering gun to remove all the solder then resolder the joint.

4.4.2. Soldering stranded wire

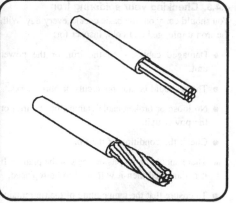

- Before connecting stranded wire to a connector it must first be stripped, twisted and tinned.

- Use pliers to twist the strands or use the method described in the section on insulation removal, using the insulation stub.

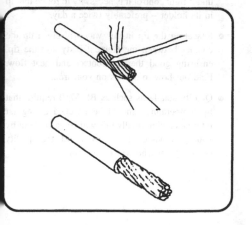

- Steady the wire and apply a light touch of solder to the strands.

- The tinning should stop just short of the insulation.
- The outline of the strands should be visible.

4. SOLDERING AND TERMINATION

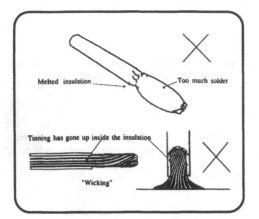

Melted insulation

Too much solder

Tinning has gone up inside the insulation

"Wicking"

- Too much solder and the wire is impossible to form.
- The insulation is also melted into the solder causing contamination.

- This view shows *wicking* and is caused by the solder running up inside the insulation. This stops the wire being flexible at the joint and it will probably break off after a time.

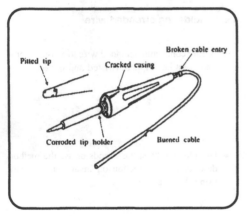

Pitted tip

Cracked casing

Broken cable entry

Corroded tip holder

Burned cable

4.4.3. Checking your soldering iron

You should carry out the basic checks every day. With the iron unplugged and cold, inspect for:

- Damaged cables – on the iron or the power unit.
- There should be no splits, cuts or burn marks.
- No loose or broken cable clamps, on the iron or the power unit.
- Check the condition of the tip.
- There should be no pits or holes in the plating. If it is damaged then it will need to be replaced.
- To ensure that the temperature of the tip remains under tight control it is necessary to rotate the tip in its holder – preferably twice a day.
- Reseating the tip in this way will clean up the contact between the element body and the tip, ensuring good thermal contact and heat flow. Find out how to do this on your iron.
- Quality standards such as BS 5750 require that tip temperatures should be checked at regular intervals and normally this will be carried out by your supervisor or a member of the quality control team who will keep records.

4. SOLDERING AND TERMINATION

Summary

Soldering

- **Remember.** Soldering takes much practice.

- Allow the iron to heat up and stabilise before you use it.

- Wet the cleaning sponge.

- Make sure that the surfaces to be soldered are clean and free from grease. The solder won't take properly otherwise.

- Use non-corrosive flux-cored solder.

- Make a mechanical joint before soldering.

- Apply the iron tip to heat both parts of the joint.

- Melt the solder on the joint not the iron.

- The wire outline should still be visible under the solder.

- Check your iron daily for signs of damage – when the iron is cold!

- Clean plated tips on a wet sponge only.

- Make sure the sponge is *kept* wet.

- Never file plated tips.

- Always replace the iron in its stand after use.

- Rotate the tip at least once a day if in constant use.

4.4.4. Safety aspects of soldering

- Molten solder can easily burn flesh and cause serious damage to eyes should there be direct contact.

- The soldering tip will burn skin and clothing.

- Keep the iron in an enclosing holder when not being used.

- Always use the damp sponge to clean the tip – *never flick the iron.*

- Protective clothing, when instructed, should be worn.

- Find out where the First Aid box is, including eyebaths, which should be available in the workplace.

- Most irons are 12 V or 24 V powered and are therefore relatively safe from the electric shock point of view.

- Mains-powered types must be earthed to the line earth and checked frequently for damage likely to cause a shock hazard.

4. SOLDERING AND TERMINATION

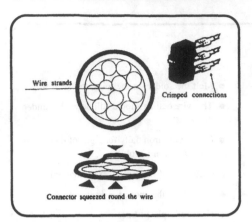

Wire strands

Crimped connections

Connector squeezed round the wire

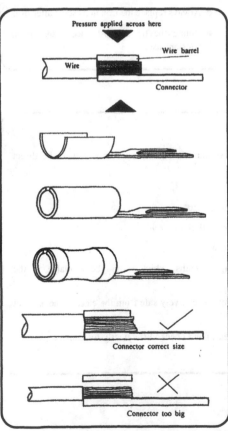

Pressure applied across here

Wire barrel

Wire

Connector

Connector correct size

Connector too big

4.5. Crimped joints

The majority of wire terminations used in control panel assembly are made with crimp connectors. The main reason for this is that they are easy and quick to produce.

Crimping simply means that the conductor is placed into a special *crimp connector* which is then compressed around it with the use of a *crimping tool*. See Section 4.5.2.

4.5.1. Crimp connectors

The crimp fitting end of the connector has a wire barrel of a suitable diameter to take the conductor. It is this part that is compressed by the crimp tool.

- The wire barrel may be open.

- Or closed.

- It may be insulated.

- The conductor should be a snug fit in the barrel.

4. SOLDERING AND TERMINATION

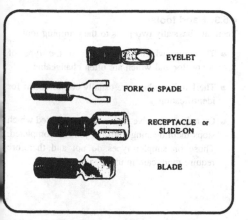

EYELET

FORK or SPADE

RECEPTACLE or SLIDE-ON

BLADE

The actual connector can have any one of a wide variety of shapes determined by the requirements of the job.

- Here are some commonly used single wire crimp connectors. All are insulated in these examples.

4.5.2. Crimping tools

The purpose of the crimp tool is to correctly apply pressure to the wire barrel to trap the conductor tightly so that it cannot be pulled out under normal circumstances. At the same time it must not be so tight as to cause strands or the connector to break.

Crimping tools may be operated in various ways dependent not only on the size of the conductor but sometimes on the total number of crimps that will be needed. However, they are all similar in operation.

- Hand-operated. Used for light duty work – smaller conductors and small quantities. These are described here in detail.

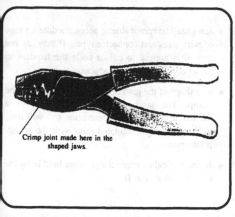

Crimp joint made here in the shaped jaws.

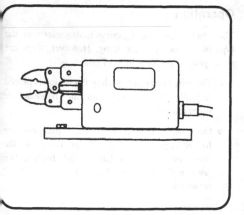

- Power-operated. These can be powered by compressed air, electric or hydraulics. Generally they are bench-mounted but there are hand-held types.

Manufacturer's instructions should be followed carefully.

SAFETY!

Take care when using power crimpers. Guards should be fitted.

4. SOLDERING AND TERMINATION

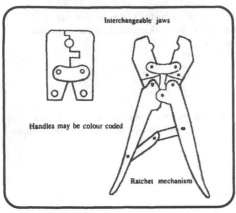

Interchangeable jaws

Handles may be colour coded

Ratchet mechanism

4.5.3. Hand tools

There are basically two parts to the crimping tool.

- The jaws which are special to the type of connector and which are often changeable.

- The handles which are usually colour coded for identification.

- On more expensive tools a ratchet is fitted which stops them opening until the joint is completed. Those on simpler types do not and therefore require more care in use.

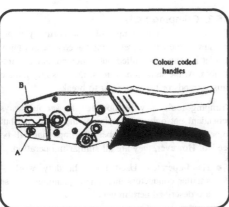

Colour coded handles

B

A

- Jaws may be removable to accommodate a range of wire sizes and connector types. If they are not then it is normal to colour code the handles to avoid confusion on a production line.

- The shape of the jaws determines the shape of the crimp. The jaws therefore are special to a particular type of crimp connector and will only give a correctly terminated joint with the appropriate parts.

- In this typical example the jaws are held in by the two screws A and B.

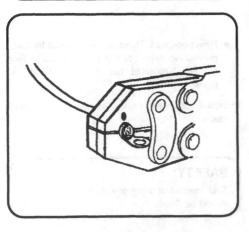

Operation

The actual detail of using crimp tools varies with the type of crimper you are using. However, there are some general points worthy of note.

- The wire barrel of the crimp connector is placed centrally in the jaws and the handles are squeezed together.

- Once the crimp has been made the jaws are locked in position by the ratchet. To release the jaws you squeeze the handles still further. The jaws will open and the joint may then be removed.

4. SOLDERING AND TERMINATION

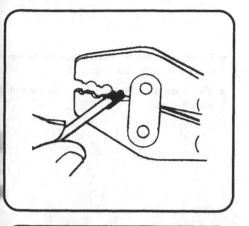

- Where no ratchet is fitted you have to gauge how hard to squeeze the handles to obtain a good crimp. This is learned by experience and has to be found by trial and error.

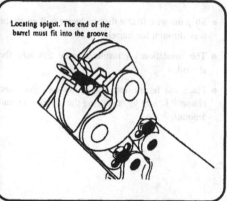

Locating spigot. The end of the barrel must fit into the groove

- Some have a locating marker – in this case a spigot – to ensure the correct location of the connector.

SAFETY!

When using a hand tool which has a ratchet mechanism in the handle, take care not to trap a finger as the operating cycle of the tool is not reversible. In other words, once the handles are squeezed together the jaws can only be opened by applying further pressure to the handles.

4.5.4. Bootlace ferrules

These are special connectors used extensively for terminating wires to be connected to screw terminals such as those found on relays and contactors.

- They come in several sizes, with each size having a different colour.

- Uninsulated versions are also available.

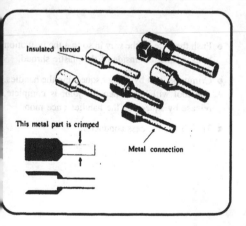

Insulated shroud

This metal part is crimped

Metal connection

4. SOLDERING AND TERMINATION

| | | (Uninsulated ferrules) |
Wire size	Colour	Internal diameter
0.5 mm^2	White	1.0 mm
0.75 mm^2	Blue	1.4 mm
1.0 mm^2	Red	1.6 mm
1.5 mm^2	Black	1.8mm
2.5 mm^2	Grey	2.3 mm
4.0 mm^2	Orange	2.8 mm

● The shroud is colour-coded to show the recommended conductor size to be used.

● Always use the correct size ferrule for the wire you are using.

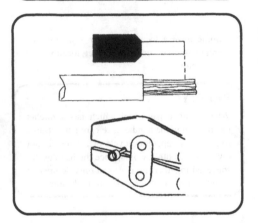

● Strip the wire so that the conductor will go all the way through the barrel.

● The insulation on insulated wire fits into the shroud.

● Place the ferrule into the crimp tool jaws and clamp it lightly by squeezing the handles a small amount.

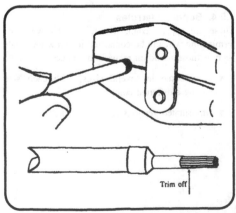

Trim off

● Push the wire all the way in so that the insulation butts against the inside of the plastic shroud.

● Crimp the joint by further squeezing the handles. The tool will lock when the joint is complete, release by squeezing the handles once more.

● Trim off the excess conductor.

4. SOLDERING AND TERMINATION

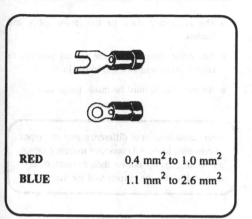

RED   0.4 mm^2 to 1.0 mm^2

BLUE   1.1 mm^2 to 2.6 mm^2

4.5.5. Insulated eyelets and spades

- These are used to terminate wires which will be fixed under a screw.

- They are also colour coded by wire size.

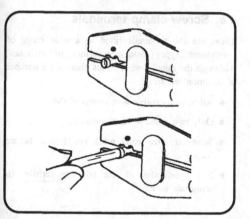

- Strip the wire to give the correct amount of exposed conductor.

- Place the connector into the crimp tool and clamp it lightly.

- Push the wire into the connector until the insulation butts against the barrel.

- Crimp the joint as before.

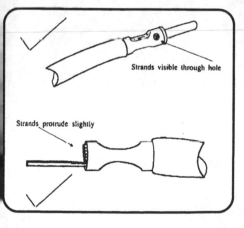

Strands visible through hole

Strands protrude slightly

4.5.6. Inspection

Most blind connectors will have some way of inspecting the wire after crimping.

- This may be a hole – found in multipole inserts. The wire strands must be visible through the hole.

- On others like the insulated eyelets, the conductor should protrude through the barrel so that it is level with the connector insulation.

4. SOLDERING AND TERMINATION

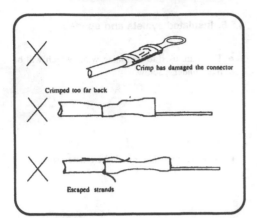

Crimp has damaged the connector

Crimped too far back

Escaped strands

- The connector must be free from splits and flashes.
- The crimp must be on the correct position to ensure maximum strength to the joint.
- All the strands must be inside the joint.

There are a number of different makes and types of crimping tools. All connector makers produce a matching crimp tool for their connectors. It is essential to use the correct tool for the job.

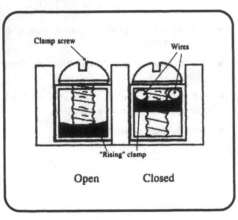

Clamp screw

Wires

"Rising" clamp

Open   Closed

4.6. Screw clamp terminals

These are the terminals fitted to a wide range of component types from contactors to switches. Although the detail design varies, there are a number of common features.

- All have a captive wire clamp washer.
- Only two wires to each connector.
- Stranded wire ends must be twisted before fitting.
- It is preferable to use bootlace ferrules to terminate wires.

4. SOLDERING AND TERMINATION

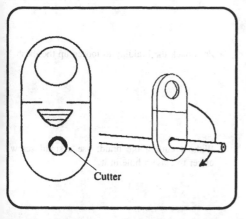

Cutter

4.7. Terminating coaxial cable

4.7.1. Stripping

The recommended method is to use one of the coaxial cable strippers currently available. The operating instructions vary according to type.

- With this tool the cable is passed through the hole after lifting the top half to open up the cutter.

- Push the top down to cut the insulation then rotate to cut it all the way round.

- Pull off the stripper and the insulation stub.

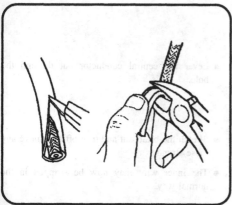

Another way using a sharp knife and wire cutters:

- Rest the cable on the workbench.

- The outer cover can be stripped back using a sharp knife to make a slit along its length.

- Take care not to damage the inner screening.

- Peel the cover back and trim off with side cutters.

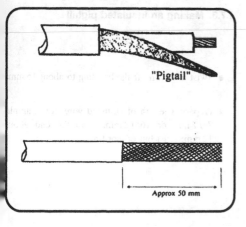

"Pigtail"

Approx 50 mm

4.7.2. Making a 'pigtail'

- This is a way of separating the braid and inner conductor before making any solder connections.

- The braiding must not be soldered while it is still on the central insulation.

- Strip about 50 mm off the outer insulation.

41

4. SOLDERING AND TERMINATION

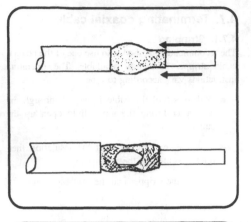

- Push back the braiding to loosen up the mesh.

- Without cutting the braid, use a small screwdriver to tease a hole in it.

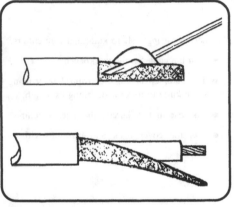

- Lever the central conductor out through the hole.

- Stretch the braid out and trim off to remove any loose strands.
- The inner wire may now be stripped in the normal way.

4.7.3. Making an insulated pigtail

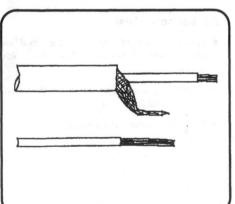

- Twist and trim off the braiding to about 15 mm.

- Prepare a length of stranded wire, for example 7/0.2 mm or 16/0.2 mm. Strip the end about 12 mm; twist but do not tin.

4. SOLDERING AND TERMINATION

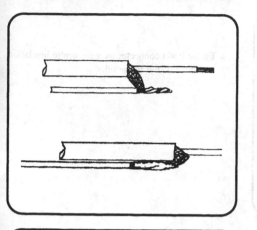

- Twist the braiding and wire together.
- Solder the joint and trim off to 8 mm long.

- Fold the connection back over the outer cover.

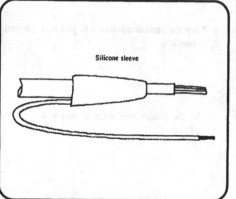

Silicone sleeve

- Fit a silicone rubber sleeve to cover the joint.

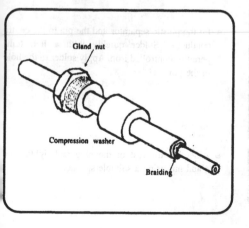

Gland nut

Compression washer

Braiding

4.7.4. Fitting a BNC coaxial plug

To terminate coaxial cable with a standard BNC plug:

- Strip off sufficient length of the outer cover and cut off the braid level with the new end of the outer cover.
- Fit the gland nut and plastic compression washer over the outer covering.

43

4. SOLDERING AND TERMINATION

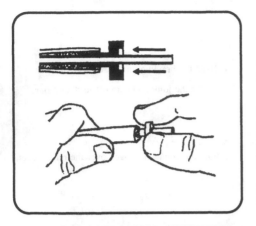

- Fit the braid connector over the centre insulation and push it into the braid.

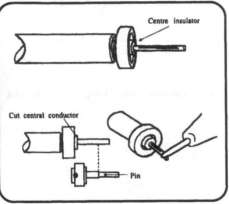

- Strip the central insulation to just past the braid connector.

- Cut the centre conductor to length and tin the end.

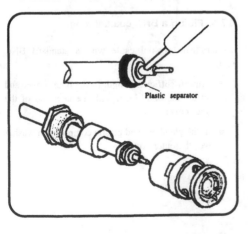

- Fit the plastic separator and the pin to the centre conductor. Solder quickly with a hot, temperature-controlled iron. Apply solder to the hole in the pin.

- Assemble the rest of the plug and tighten the gland nut using a suitable spanner.

44

4. SOLDERING AND TERMINATION

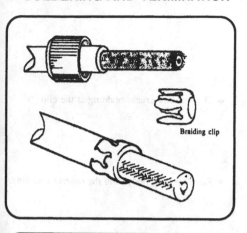

Braiding clip

4.7.5. Fitting a push-fit coaxial plug

- Strip off sufficient length of outer covering.

- Fit the plug cover nut.

- Push the braiding clip over the braiding, making sure that it is hard up against the insulation then squeeze the jaws lightly to grip the insulation.

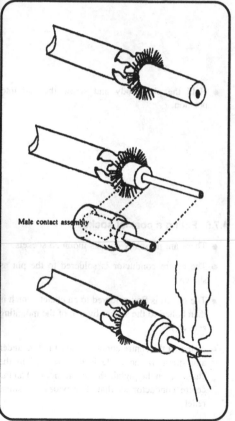

Male contact assembly

- Comb out the braiding and trim down to about 10 mm long. Fold it back over the clip.

- Strip the centre insulation to the required length and tin the conductor lightly with the solder.

- Push on the male contact assembly so that it is hard up against the braiding. Hold it there and solder the centre conductor in place.

4. SOLDERING AND TERMINATION

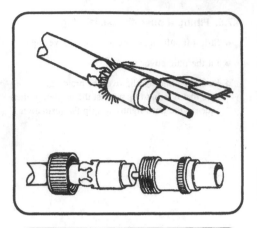

- Trim off any excess braiding at the clip.

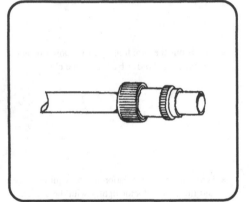

- Place the plug body onto the contact assembly.

- Hold the plug body and screw the cap into position.

4.7.6. Fitting a coaxial socket

- These are normally chassis-mounted sockets.

- The centre conductor is soldered to the pin as shown.

- The pigtail is first soldered to an eyelet, which is then bolted to the socket by one of the mounting screws.

- Note that the centre conductor must not be under any strain. If the cable is not clamped to the chassis then the pigtail should be shorter than the centre conductor so that it provides the stress relief.

5. CABLE FORMING

5.1. Cableforms

A cableform is where a number of individual wires, which may be of different sizes and types, are bound together to form a single cable run. Alternative names are *cable harness* or *wiring loom*.

Cableforms are often made up as a separate item along with other components for the equipment in which they will be installed and the following information is usually provided:

- Wiring schedule.
- Cableform template.
- Run-out sheet or table.

5.1.1. Wiring schedule

This gives details of the wires used in the cableform. The exact layout will depend on your company but will normally include:

- Type of wire – number of strands – size of strands – insulation – colour;
- Ident marker;
- Length;
- Stripping or termination details.

Refer to this table to make up the individual wires.

Wire No	Type	Colour	Length	Strip length
01	16/0.2	Black	800 mm	10 mm
02	16/0.2	Black	800 mm	10 mm
03	32/0.2	Red	1200mm	10 mm
04	32/0.2	Red	450.mm	10 mm

5.1.2. Template

- This is a full size plan view of the cableform. The position of forming pins as well as the position of the wire ends will be marked.
- On larger cableforms the template will be divided into zones to make both ends of a wire easier to locate.
- The template is fixed to a piece of board and used as a pattern. Forming pins or smooth nails are put in at the relevant points on the template.

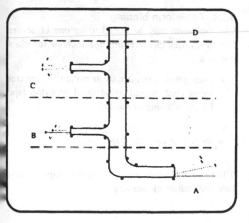

5. CABLE FORMING

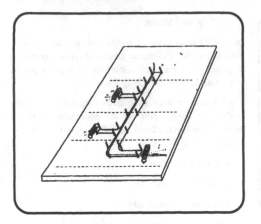

- The cableform is made by laying the wires between the connection points and following the shape made by the forming pins which keep the wires together until they are bound into a cableform.

Run-out sheet

Wire No	From	To	Colour
01	A	C	Black
02	A	C	Black
03	A	D	Red
04	A	B	Red

5.1.3. Run-out sheet

- This gives the order in which the wires are laid into the cableform and the zone location of the wire ends.

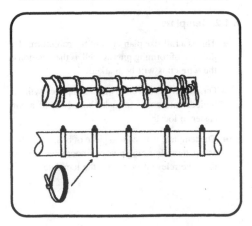

5.1.4. Cableform binding

The cableform may be bound using one of several methods. Check the cableform specification for which one to use.

- Lacing with a *continuous tie* using PVC-covered nylon cord, waxed nylon braid or nylon tape. Called *stitching* or *lacing*.

- Individual ties called *spot ties*.

There are other bindings such as spiral wrap, adhesive tape and heatshrink sleeving.

5. CABLE FORMING

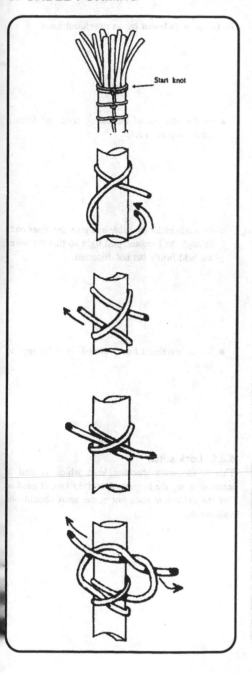

Start knot

5.2. Continuous lacing

5.2.1. Start knots
There are two variations used in the industry.

(i) Clove hitch followed by an overhand knot

- Loop about 150 mm of the cord under the cable and pass it over the long length.

- Make another loop passing the end under the first.

- Pull tight so that the cable is held firm but not distorted.

- Tie an overhand knot. Varnish may be applied later.

5. CABLE FORMING

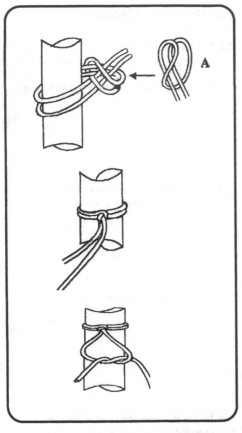

(ii) Loop tie followed by an overhand knot

- Double the end of the lacing cord and form a small loop as in A.

- Pass this under the cable and pass the other ends through the loop and pull tight so that the wires are held firmly but not distorted.

- Tie an overhand knot. Varnish may be applied later.

5.2.2. Lock stitch

This is the main stitching knot which is tied at intervals along the harness. A *locking knot* is used so that the cableform does not come apart should one knot break.

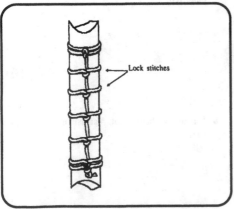

Lock stitches

5. CABLE FORMING

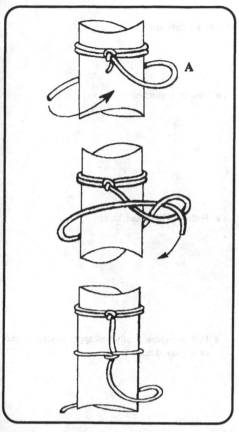

- Take the cord under the cable leaving a loop as at A.

- Hold this loop as shown and pass the cord through.

- Pull it tight as for the start knot and at the same time manoeuvre the knot to be in line with the others.

- Space the knots at about 1.5 times the cable diameter.

5.2.3. Double knot

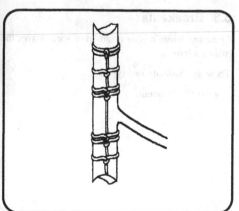

- This is simply two lock stitches tied close together for extra strength.

- Used at the finish of a lacing run and also where a number of wires leave the main cableform.

5. CABLE FORMING

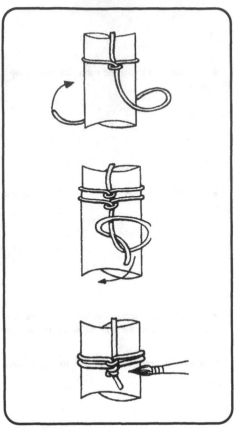

5.2.4. Finish knot

- Two lock stitches.

- Followed by a reef knot.

- Pull to tighten firmly and apply staking lacquer or approved adhesive.

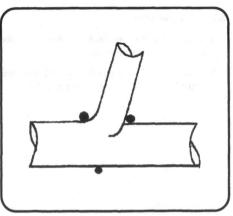

5.3. Breakouts

These are where a wire or group of wires leaves the main cableform.

There are basically two types:

- (i) 'Y' breakout.

5. CABLE FORMING

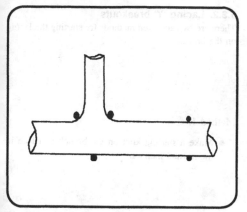

- (ii) 'T' breakout.

5.3.1. Lacing breakouts

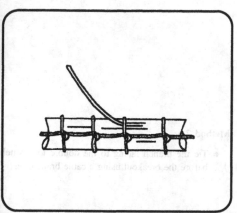

- Single wires are brought out after a lock stitch.

Where there are several wires:

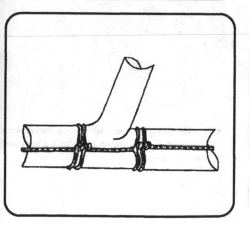

- Make a double lock stitch before and after the breakout. Continue lacing along the main cableform.

5. CABLE FORMING

5.3.2. Lacing 'Y' breakouts

There are two accepted methods for starting the lacing on the branch.

Method 1.

- Make a starting knot on the branch and lace in the normal manner.

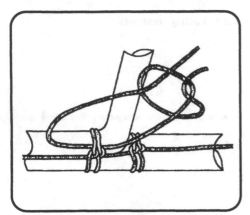

Method 2.

- Tie the branch lacing to the double lock stitch before the breakout using a cable branch tie.

- Pull the tie tight to the double stitch then make a lock stitch on the branch close to the join and continue lacing.

5. CABLE FORMING

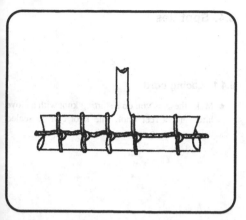

5.3.3. Lacing 'T' breakouts

There are several methods which may be used depending upon the specification.

Method 1. Where there are only a couple of wires (*less than 12 mm diameter overall*).

- Make a *single lock stitch* on both sides of the breakout.
- If the breakout is to be laced, use a starting knot and lace as normal.

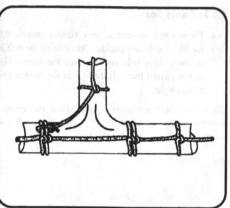

Method 2. On larger breakouts *more than 12 mm*.

- Make a double lock stitch before and after the breakout.

- Lace the branch starting with a cable branch tie.

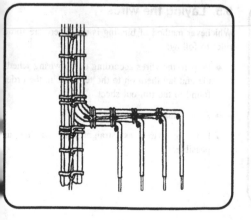

- Where the wires fan out to go to a sub-assembly or connector, use a double stitch before each wire.

5. CABLE FORMING

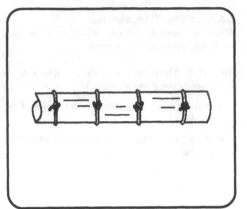

5.4. Spot ties

5.4.1. Lacing cord

- Make these as you do a starting knot with a clove hitch and a reef knot. The knot can be sealed using adhesive or varnish.

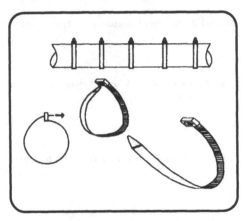

5.4.2. Cable ties

- These come in several very similar shapes. All are like a belt and buckle. One side of the belt is serrated. This side goes toward the cable. The end is passed through the eye in the buckle and pulled tight.

There are tools available which allow the correct tension to be obtained every time. Trim off the waste.

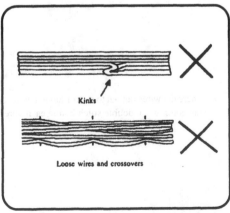

Kinks

Loose wires and crossovers

5.5. Laying the wires

Whichever method of binding is used, here are some rules to follow:

- Prepare the wires according to the wiring schedule and lay them on to the template in the order found in the run-out sheet.

- Avoid kinks in the wires.

- Lay the wires as straight and parallel as possible.

5. CABLE FORMING

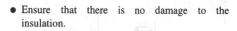

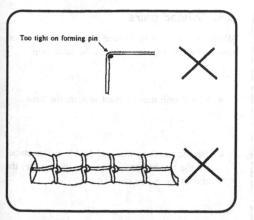

Too tight on forming pin

- Ensure that there is no damage to the insulation.

- The wires and insulation must not be damaged in any way by the binding. Damaged wires must be replaced.

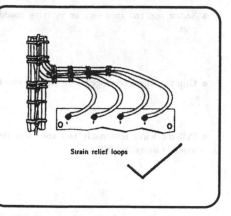

Strain relief loops

- Where wires exit from the cable form they must not be under strain. Leave enough wire to give a neat loop and avoid undue cross-overs.

- Mains or other power wires and cables **must not** be included in a cableform with any other wires.

Cableform diameter	Tape size
Up to 5 mm	0.9 mm x 0.15 mm
5.1 mm to 10 mm	1.25 mm x 0,2 mm
10.1 mm to 20 mm	1.5 mm x 0.3 mm
Over 20 mm	2.25 mm x 0.3 mm

- Use a lacing tape of the correct size, according to this table.

- The spacing between ties or lock stitches should be about 1.5 times the cable diameter, but no more than twice.

57

5. CABLE FORMING

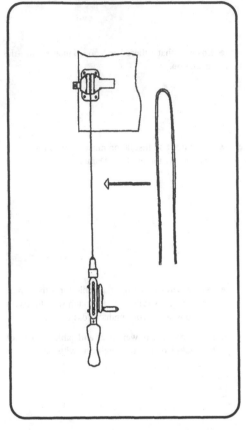

5.6. Twisted pairs

Wires may have to be twisted together for electrical reasons before they are laid into the cableform.

- A hand drill may be used to form the twists.

- Use one length of wire and double it. Remember that the finished twisted pair is shorter than the straight piece you start with.

- Secure one end in a vice or by tying round a post.

- Grip the other ends together in the drill chuck.

- Pull the wires reasonably taut and twist them together using the drill.

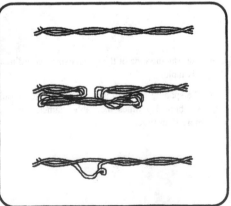

- Twist evenly and neatly.

- Don't twist so tight that the wires kink or loop back on themselves.

- Avoid gaps and loops between wires.

5. CABLE FORMING

Length of one full twist		
Wire size	Minimum	Maximum
7/0.2	14 mm	15 mm
14/0.2	19 mm	29 mm
19/0.2	22 mm	33 mm
23/0.2	25 mm	37 mm

• The length of twist or the number of twists is determined by the wire diameter.

• This table gives a guide. If one wire is thicker then this diameter determines the length of twist.

Summary

• Mains and other power wiring must be formed into a separate cable form from all other wires.

• Power wiring must conform to the IEE wiring regulations.

• There must be no strain on any wires at junctions or breakouts.

• There must be no kinks in wires.

• Leave enough wire length at the ends so that a strain relief loop can be provided where needed.

• In general there should be enough wire length for the joint to be made at least twice, including 10 mm stripping length.

Cableform routing

When laying the cableform into the equipment avoid:

• Heat-generating parts.

• Sharp edges on hardware.

• Moving parts.

• Covering service parts that need access.

5. CABLE FORMING

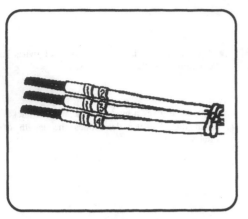

5.7. Cable markers

Cable or wire markers are used simply to identify wires, especially in multiway cables or wiring harnesses. Both ends are marked with the same numbers to be used.

- Often these numbers will be the same as those on the connector to which they will be connected. In any case the wiring drawing or run-out sheet will give the wire numbers to be used.

- Most have numbers printed on as well as being coloured, although there are several types which are coloured only. Some are wrapped round the wire and are adhesive, while others are like small sleeves which slip over the insulation.

BLACK	0
BROWN	1
RED	2
ORANGE	3
YELLOW	4
GREEN	5
BLUE	6
VIOLET	7
GREY/SLATE	8
WHITE	9

- The colours used to represent the numbers are the same as the resistor code so there is nothing new to remember!

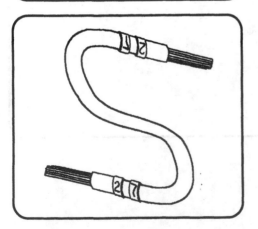

- The markers are placed so that the number is read from the joint as illustrated. This example shows wire number 27.

5. CABLE FORMING

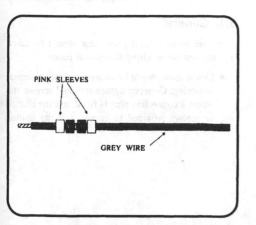

PINK SLEEVES

GREY WIRE

- When the markers are the same colour as the insulation then unmarked sleeves of a contrasting colour, usually pink, are placed on either side of them to highlight their presence.

- The example illustrates this with the number 88 (grey/grey), assigned to a wire with grey insulation.

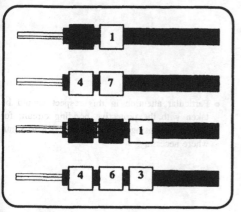

- When the number of the last wire to be marked is more than one digit, additional zeros are added in front of the lower numbers to give them the same number of digits as the last wire.

- For instance, if the last wire is between 10 and 99, then a 'zero' or black marker is placed before all single digit numbers. This makes 1 become 01, 2 become 02 and so on.

- Similarly with a last wire number of between 100 and 999, two zeros are added so that 1 becomes 001, 11 becomes 011, and so on.

6. CONNECTIONS AND ROUTING

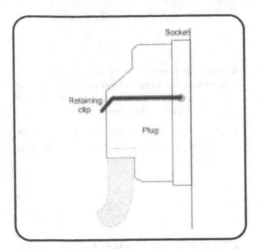

6.1. General

These are some general points that should be taken into account when wiring the control panel.

- Connections should be secured against accidental loosening. Correctly tighten terminal screws and where a connecting plug is fitted, use the clamps or screws provided to secure it to its mating socket.

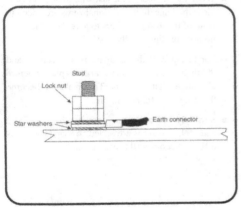

- Particular attention in this respect should be taken with the protective bonding circuit, for example by using star washers and a lock nut where necessary.

6. CONNECTIONS AND ROUTING

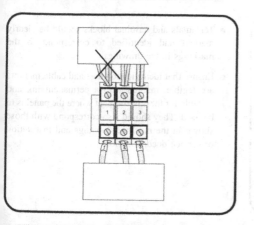

- Two or more conductors may only be connected to a terminal that is designed for the purpose. The majority of connecting blocks will only take one or two conductors. Don't force in any more.

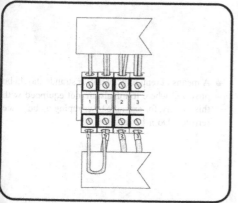

- Add an additional terminal and connect it to the other by a link laid in the cable trucking to gain an extra connection point.

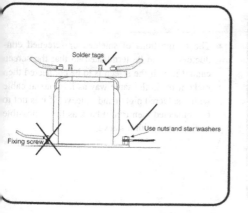

- Soldered connections should be made only to terminals suitable for that purpose. Transformers may be fitted with turret tags suitable for soldering and printed circuit board assemblies may have solder pins.

6. CONNECTIONS AND ROUTING

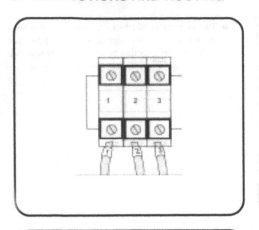

● Terminals and terminal blocks should be clearly marked and identified to correspond to the markings in the drawings.

● Ensure that identification tags and cable markers are legible, marked with a permanent ink and suitable for the environment where the panel is to be used. They should also correspond with those shown in the machine drawings and instruction or service documentation.

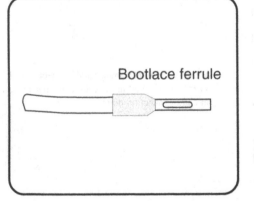

Bootlace ferrule

● A means of retaining conductor strands should be provided where terminals are not equipped with this facility, for example by crimping on bootlace ferrules. Do not use solder.

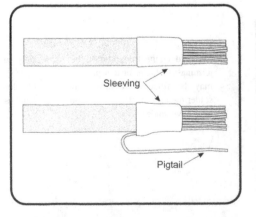

Sleeving

Pigtail

● The terminations of shielded or screened conductors should be terminated so that the screen cannot fray. If the screen is to be connected then make it off in the same way as for coaxial cable with a soldered pigtail and a sleeve. If it is not to be connected then trim it back as far as possible and cover it with a sleeve.

6. CONNECTIONS AND ROUTING

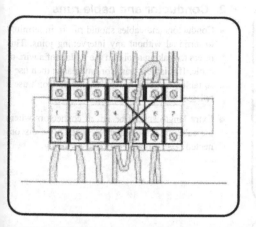

● Terminal blocks should be mounted and wired so that the internal and external wiring does not cross over the terminals.

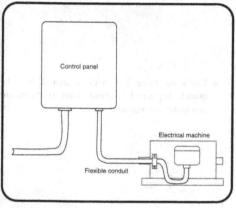

● Flexible conduits and cables should be installed in such a way that liquids can drain away from fittings and terminations.

6. CONNECTIONS AND ROUTING

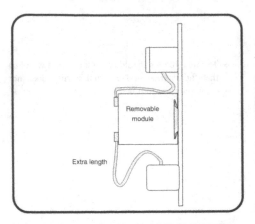

6.2. Conductor and cable runs

- Conductors and cables should run from terminal to terminal without any intervening joins. This refers to making a joint in the middle of a wire or cable. If it is necessary for any reason then use a suitable connector or terminal block. Don't use a twisted and soldered joint.

- Extra length should be left at connectors where the cable or cable assembly needs to be disconnected during maintenance or servicing.

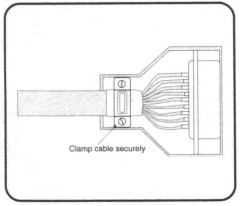

- Multicore cable terminations should be adequately supported to avoid undue strain on the conductor terminations.

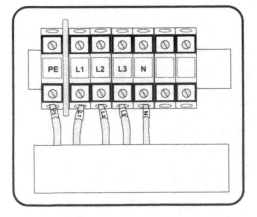

- The protective conductor should as far as is possible be routed close to the associated live conductors to avoid undue loop resistance.

6. CONNECTIONS AND ROUTING

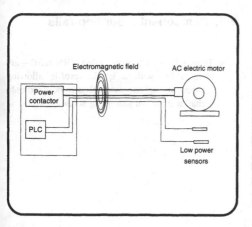

Electromagnetic field

AC electric motor

Power contactor

PLC

Low power sensors

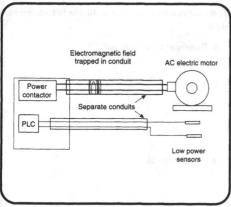

Electromagnetic field trapped in conduit

AC electric motor

Power contactor

Separate conduits

PLC

Low power sensors

6.3. Conductors of different circuits

This refers to wires and cables that are in the same enclosure but are connected to different parts of the system, for example power wiring that could be carrying high currents at 415 volts. Signal wires that may be connected to sensors and to the input terminals of a programmable controller and therefore carrying only low currents at 5 to 24 volts.

When a conductor is carrying current, an electromagnetic field is produced. This is more pronounced when the power is high such as may be the case for a powerful electric a.c. motor. This field can cause a voltage to be generated in other conductors nearby. It is possible for this so-called interference voltage to cause another circuit to react, causing a malfunction.

When the current is switched on or off, the electromagnetic field increases and decreases, rapidly causing, in effect, a radio signal. The effect is similar to the crackle that can sometimes be heard on the radio or television when something like a fridge switches on and off. This radiated signal can be picked up by the other wires in the system and cause interference to the normal working voltages in the system. This is known as Electromagnetic Interference or EMI. The Electromagnetic Compatibility (EMC) Regulations require that these effects are minimised.

In many cases, it is the layout of the wiring that is critical to avoid interference and this will have been worked out by the designer. This aspect of wiring concerns the EMC regulations.

- The layout of such wiring should be specified by the designer and must be adhered to.

- Where the circuits work at different voltages, the conductors must be separated by suitable barriers or all the wires insulated for the highest voltage to which any conductor may be subjected.

Circuits which are NOT switched off by the supply disconnecting device should be either separated physically from other wiring and/or distinguished by colour so that they can easily be identified as being LIVE when the disconnecting device is in the off or open position.

A lamp inside an enclosure provided for use during maintenance is an example of such a circuit. The control panel may be isolated but the lamp will require power so that the engineer can see while working on it.

7. HARDWARE

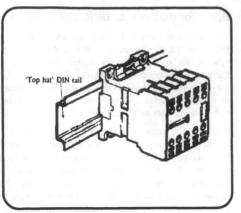

'Top hat' DIN rail

7.1. Component mounting rails

- These rails – sometimes called 'DIN rails' – are metal strips with a special profile allowing components and sub-assemblies to be fixed onto a chassis plate without using screws.

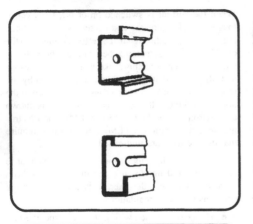

There are two basic profiles available in two common sizes:

- Symmetrical or 'top hat'.

- Asymmetrical.

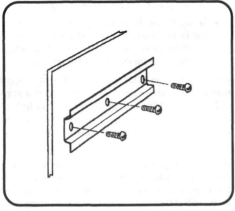

- These are cut to the required length and then screwed or bolted to the chassis before any wiring begins.

7. HARDWARE

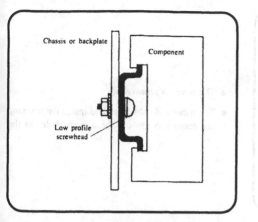

- To avoid fouling the underneath of components, use screws with low profile heads.

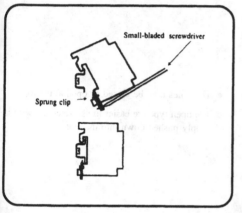

Components are then clipped on. The actual detail depends on the profile. The method used for 'top hat' is shown here.

- Locate the top of the rail in the top groove at the back of the component.

- Rotate it downwards to cause the spring clip to retract and snap into place behind the rail.

- There is a slot in the spring clip so that it can be retracted using a small screwdriver.

- Although this is mainly intended to be used when the component is removed, it may be necessary to use it when mounting the component if it appears to need too much force.

7.2. Plastic trunking

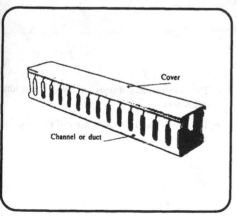

- This is one form of cable ducting and used to carry the wiring between components. It provides protection while keeping the wires and cables neat.

7. HARDWARE

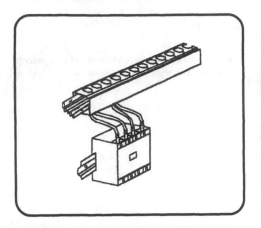

- The cover is removable.
- The wires and cables are laid inside the trunking and leads brought out through the holes in the side.

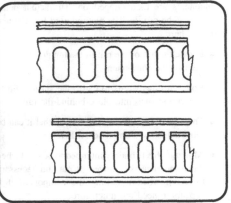

- The holes may be closed or open at the top.
- The open type are easier to use since the wire is simply pushed down into the slot.

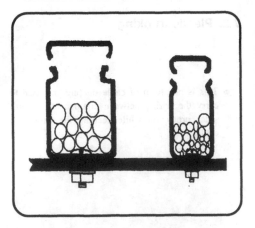

- There are a variety of sizes in terms of both width and height.
- The wires should not more than half fill the cable duct.

7. HARDWARE

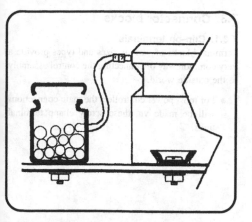

- Plan the cable run so that there is the minimum of cross-overs.

- However, still leave enough spare wire to make the connection at least twice.

- Loop the wires neatly from the component to the trunking.

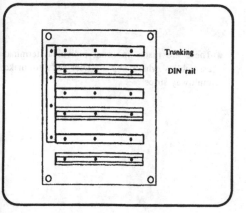

Trunking

DIN rail

- A chassis before components or wiring may look like this, with the din rails and trunking fitted.

- The next stage is to clip the components to the rail and wire them together.

7. HARDWARE

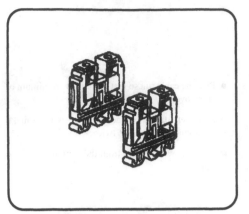

7.3. Connector blocks

7.3.1. Clip-on terminals

Terminal blocks of various sizes and types provide a very common way of connecting the control assembly to the outside world.

- For most power controllers the main connections will be made via these screw clamp terminal units.

- These are made up using individual terminal assemblies which clip to DIN rails to make multiway strips.

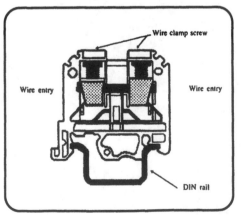

Wire clamp screw

Wire entry

Wire entry

DIN rail

- The terminal is specified in terms of the cross-sectional area of wire it will accept. This varies from $1.5\,mm^2$ upwards.

- The most common way of terminating wires is the screw clamp with no more than two wires per terminal.

- The wire sizes are specified in data tables for plain wire ends and for those with ferrules on.

- Different colours may be used to aid identification of groups of functions.

7. HARDWARE

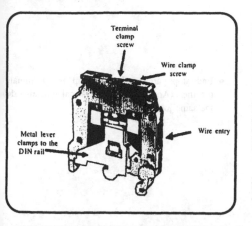

Terminal clamp screw

Wire clamp screw

Wire entry

Metal lever clamps to the DIN rail

- An earth terminal – usually green or green yellow – clamps to the rail to ensure the case and chassis are earthed.

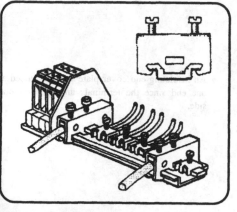

- Where multiple earths are used it is often necessary to use extra commoning clamps.

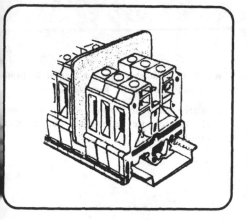

- Insulated separators can be used to further isolate high voltage connections from the others.

7. HARDWARE

● End stops are used to clamp the terminals together. (An earth clamp terminal will also do the same job.)

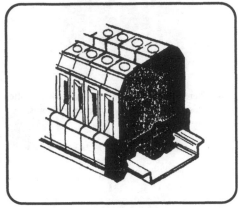

● An insulating end cover plate will be needed at one end since the terminals are open on one side.

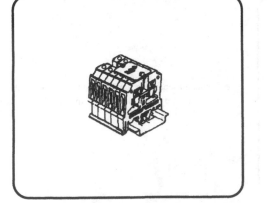

● Identifying numbers can be clipped to them, normally matching the wire indents.

● Warning covers to minimise shock risk should also be used to cover terminals carrying more than 100 V.

7. HARDWARE

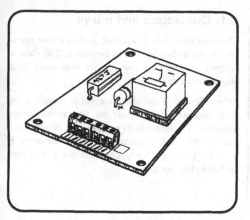

7.4. Screw terminals

7.4.1. Barrier strips

- These are used mainly on sub-assemblies to allow them to be connected into the system.

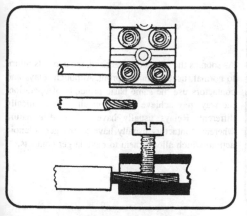

- Others have screw terminals at both sides and can be used to join wires or as a substitute for the snap together terminals in small low power assemblies.

- The simpler type have no clamping plate.

- The wires should be stripped and twisted but not tinned before inserting under the screw heads.

- Trim off so that the conductor does not go more than half way through the connector.

- Single strand wire should be folded back to give additional thickness.

- Avoid overtightening the screw because this can crush the strands and give a weak connection.

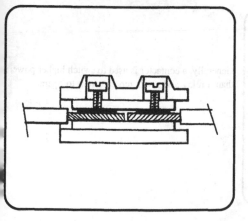

Barrier strips with clamping plates provide a secure and electrically sound termination.

8. COMPONENTS (ACTIVE)

8.1. Contactors and relays

These are mechanical switching devices whose operation is controlled by an electromagnet. The electromagnet consists of a coil of wire with many turns wound on to an iron core.

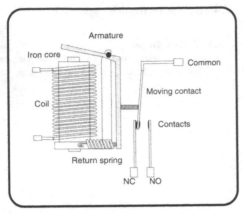

When the coil of the electromagnet is energised, the core becomes magnetised and attracts a moving armature. The armature is mechanically coupled to a set of electrical contacts. When the armature is attracted to the electromagnet, these contacts operate and complete the circuit.

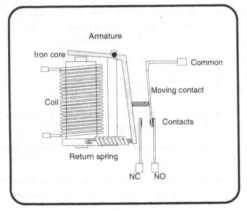

As soon as the coil is de-energised, the contacts return to normal, usually under spring. Although relays and contactors use the same basic principle of operation, the way they achieve the end result is mechanically different. Relays usually have a hinged armature whereas contactors usually have a stronger solenoid action, which allows them to have larger contacts.

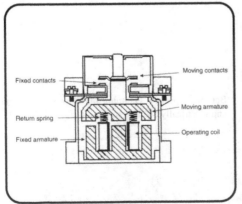

Generally, a contactor is used to switch higher powers than a relay and needs more current to operate.

8. COMPONENTS (ACTIVE)

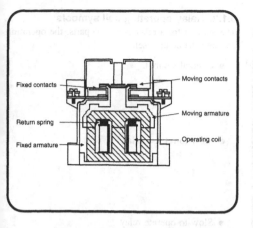

Fixed contacts

Return spring

Fixed armature

Moving contacts

Moving armature

Operating coil

Control relays use the same principle as contactors and look similar but are usually smaller. They are intended for use in the control circuit and their contacts have a lower power rating than those of a contactor.

8.1.1. Electrical specification

Electrically, contactors consist of two main parts, the operating coil and the switching contacts. A contactor will have a number of contacts (or poles), usually three normally open contacts for power switching and a set of auxiliary contacts for use at lower current in the control circuit.

Their basic electrical specifications are mainly concerned with:

- the voltage required to operate the coil;
- whether the coil needs AC or DC;
- the current-carrying capacity of the contacts;
- the maximum voltage the contacts can switch.

The type of operation they will be used for further complicates the specification – for example, how often they will make and break in an hour and whether the load is inductive (an electric motor) or resistive (a heater element).

The choice of contactor depends upon:

- the type of voltage and mains supply;
- the load power;
- the load characteristics;
- the duty requirements.

These are combined into several categories. Briefly they are as follows:

For AC loads:

AC1 – resistive load switching. Least severe conditions.

AC2 – slip ring motor control switching.

AC3 – squirrel cage motor starting and breaking during normal running.

AC4 – as for AC3 but with higher operating frequency and also where the contactor may be required to break the motor starting current. Most severe conditions.

For DC loads:

DC1 – mainly resistive loads. Least severe conditions.

DC2 – starting and stopping shunt motors.

DC3 – as DC2 but allowing inching and plugging control.

DC4 – starting and stopping series motors.

DC5 – as DC4 but allowing inching and plugging control functions. Most severe conditions.

The use of a contactor – or relay – that is not up to the conditions in the circuit will rapidly fail in service.

The contacts may weld or stick together causing power to be applied to a circuit after the contactor has been switched off.

Too much current can cause the contact to melt and disintegrate like a fuse.

8. COMPONENTS (ACTIVE)

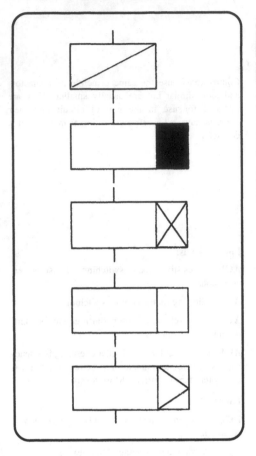

8.1.2. Relay operating coil symbols

The symbol for a relay is in two parts: the operating coil and the contact set.

- General symbol.

- Slow-to-release relay.

- Slow-to-operate relay.

- Polarised relay.

- Mechanically latched relay.

8. COMPONENTS (ACTIVE)

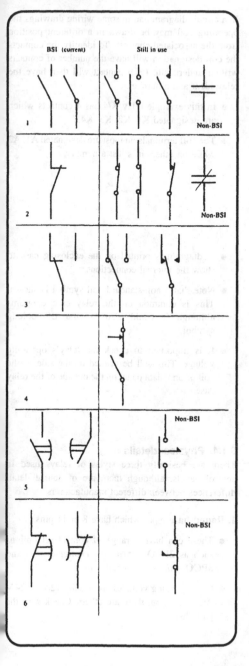

BSI (current)   Still in use

Non-BSI

Non-BSI

Non-BSI

Non-BSI

8.1.3. Relay contacts symbols

The left column shows the BSI symbols in current use. The right hand column shows other symbols in common use.

1. Make contact, normally open – N/O.

2. Break contact, normally closed – N/C.

3. Changeover (break before make) – C/O.

4. Changeover (make before break) – C/O.

5. Make, after delay.

6. Break, after delay.

8. COMPONENTS (ACTIVE)

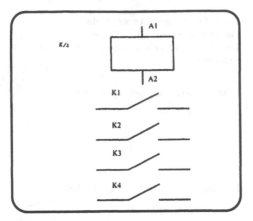

In a circuit diagram and in some wiring drawings the operating coil may be drawn in a different position from the associated contacts. To identify the contacts the coil designation will have the number of contacts written underneath. Each contact will then have the relay ident and a number.

- In this example relay K/4 has 4 contacts which are designated K1, K2, K3, K4.

- The coil terminals on most are designated A1, A2 while on others it is just a number.

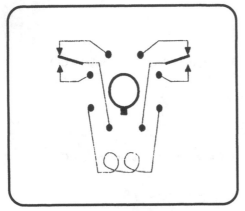

- A diagram is printed on the enclosing case to show the internal connections.

- Note that a non-standard coil symbol is shown. This is common on the relay case diagrams although it could also be the BS oblong symbol.

- It is important to check the relay's operating voltage. This will be printed on the side of the coil or on a data panel on the outside of the relay case.

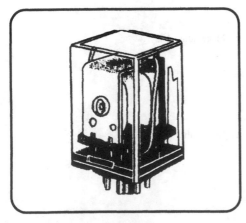

8.1.4. Physical details

There are basically three styles of relays used in control panels, although there are of course detail differences between different manufacturers.

1. **Round base** types which have 8 or 11 pins.

- These can have a range of contact set options such as **2PCO** – *two pole change-over* – and **3PCO** – *three pole change-over*.

- Coil operating voltages are usually 12 V or 24 V AC or DC, but there are others. Check with the parts list.

8. COMPONENTS (ACTIVE)

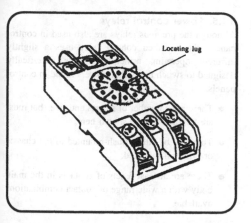

Locating lug

These plug into a base which can be bolted to the chassis or may be clipped to a DIN rail.

- The relay base has a central pin with a locating lug to provide correct orientation and to provide a reference for pin numbers.

- The connections are screw clamp type and will be numbered to correspond to the relay base.

- There will be a spring clip to hold the relay in place.

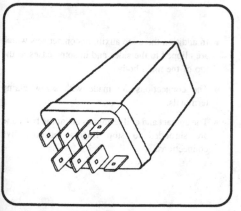

2. Square base with flat pins.

- Internally similar to the previous relay.

- Often bolted directly to the chassis or to a relay mounting plate.

- The connections are made using receptacle-type connectors.

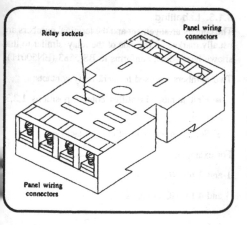

Relay sockets

Panel wiring connectors

Panel wiring connectors

- Matching DIN rail/chassis sockets can be used.

- These have screw clamp wire connections.

Note that there is a wide selection of similar relay types available which can be confirmed by a glance through the wholesale catalogues currently available.

8. COMPONENTS (ACTIVE)

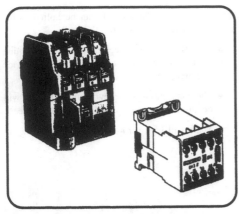

8.1.5. Power control relays

Although the previous relays are also used in control panels, the so-called control relay uses a slightly different operating principle and is specifically designed to switch the higher powers found in control panels.

- There are a number of shapes and sizes but most are similar to those shown here.

- They can be either flush-mounted to the chassis or DIN rail-mounted.

- There are at least 3 sets of contacts in the main body with a wide range of contact combinations available.

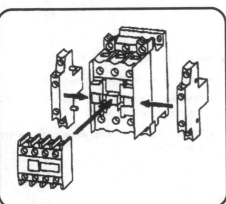

- In addition there are auxiliary contact sets which are clipped to the sides and in some cases to the top of the main body.

- The connections are made with screw clamp terminals.

- The contact and coil terminals are at the front and are shrouded to stop fingers touching live connections.

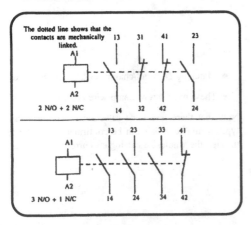

8.1.6. Labelling

The contact arrangement and the terminal numbers are usually marked on the side of the relay, similar to that shown here which conforms to BS 5583 (EN50011).

Two numbers are used to mark relay contacts:

- First number identifies contact positions 1,2,3, etc.;

- Second number identifies contact type.

For example:

1 and 2 for NC contacts;

3 and 4 for NO contacts.

8. COMPONENTS (ACTIVE)

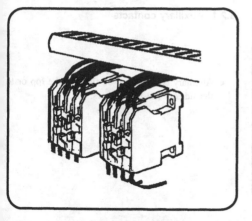

8.2. Contactors

These are an even larger form of relay designed to switch high power to motors, lamps and similar electrical power devices.

- Like control relays there is a selection of shapes and sizes. The larger the power they switch, the larger they are physically.

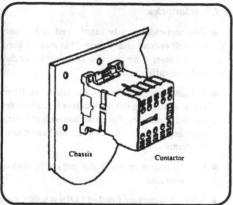

Chassis   Contactor

- They can be bolted flush with the chassis.

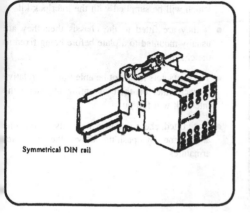

Symmetrical DIN rail

- Alternatively, they can be mounted to a DIN rail.

8. COMPONENTS (ACTIVE)

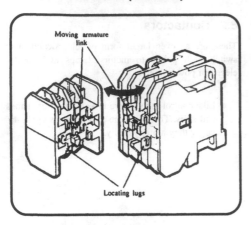

Moving armature link

Locating lugs

8.2.1. Auxiliary contacts

● Auxiliary contacts can be fitted to the top or to the sides of most contactors.

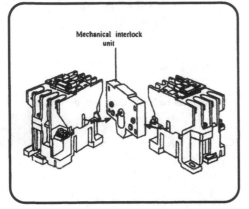

Mechanical interlock unit

8.2.2. Interlocks

● Two contactors may be interlocked so that only one will operate at any time. This may be used, for instance, when the two contactors switch a motor in different directions.

● The actual detail of fitting these interlocks differs but in general the interlock unit is fitted between the two contactors. For exact methods of fitting, read the instruction leaflet which will accompany the contactor.

● Moving spigots on either side engage in slots on each contactor.

● If the contactors are fitted to a DIN rail they must also be clipped together using the spring clips which will be supplied with the interlock kit.

● If they are fitted to the chassis then they are usually mounted to a plate before being fixed to the chassis.

● The contactors must not be able to move relative to each other on their mounting otherwise the interlock will fall out.

● Once fitted, check that they will only operate one at a time by pushing down the contactor armatures.

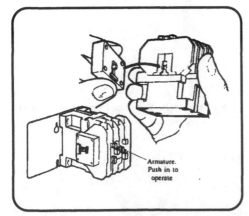

Armature. Push in to operate

8. COMPONENTS (ACTIVE)

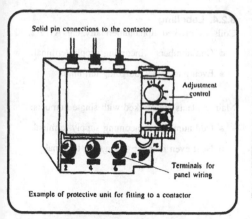

Solid pin connections to the contactor

Adjustment control

Terminals for panel wiring

Example of protective unit for fitting to a contactor

8.2.3. Protective units

Further add-on parts to a contactor-type control system include overload prevention devices.

- A protection unit may have to be fitted, e.g. a thermal overload unit.

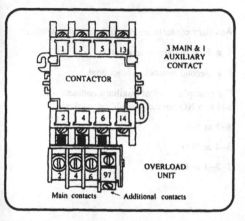

3 MAIN & 1 AUXILIARY CONTACT

CONTACTOR

OVERLOAD UNIT

Main contacts — Additional contacts

- These have three pin connectors which engage into the contactor's screw clamps.

- The overload unit has a changeover contact unit in addition to the three protected connections.

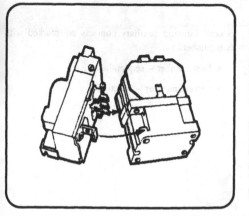

- Most also have a clip to secure them to the base of the contactor.

8. COMPONENTS (ACTIVE)

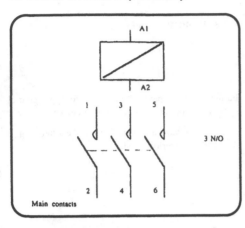

Main contacts

8.2.4. Labelling

Coils are marked alphanumerically, e.g. A1, A2.

- Odd numbers – incoming supply terminal.
- Even numbers – outgoing terminal.

Main contacts are marked with single numbers:

- Odd numbers – incoming supply terminal.
- Next even number – outgoing terminal.

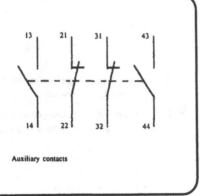

Auxiliary contacts

Auxiliary contacts are marked with two numbers:

- First number – sequential.
- Second number – functional.

For example, a set of auxiliary contacts with two NC and two NO contacts would be marked:

1–2 as N/C

3–4 as N/O

1–2–3 as C/O

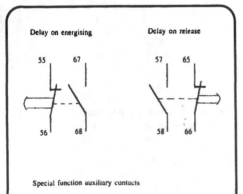

Special function auxiliary contacts

Special function auxiliary contacts are marked with two numbers:

- First number – sequential.
- Second number – functional.

8. COMPONENTS (ACTIVE)

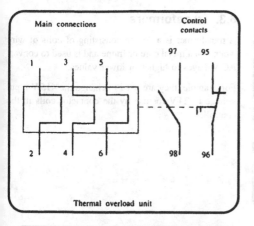

Main connections

Control contacts

97 95

1 3 5

2 4 6

98 96

Thermal overload unit

Overload relays are marked:

- Main circuit – as main contacts.
- Auxiliary contacts – two numbers.
- First number – normally 9.
- Second number – functional.

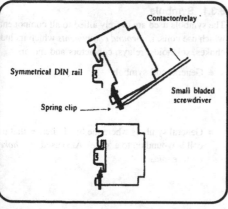

Contactor/relay

Symmetrical DIN rail

Small bladed screwdriver

Spring clip

8.2.5. DIN rail mounting

- Locate the top of the rail in the top groove at the back of the contactor or relay.
- Rotate it downwards against the lower lip of the rail which will cause the spring clip to retract and snap into place behind the rail.

There is a slot in the spring clip so that the clip can be retracted using a small screwdriver. Although this is mainly intended to be used when the component is removed, it may be necessary to use it when mounting the component if it appears to need too much force.

Summary

It is essential that the correct relay or contactor is used in a panel. The wrong electrical specification can cause a lot of damage and may impair the safety of the finished equipment.

Cross-check the following points between the parts list and component:

Coil

- Voltage.
- AC 50 Hz/60 Hz or DC.
- Power – watts.

Contacts

- Types – N/O, N/C, C/O or special function.
- Electrical ratings – voltage and current.
- Maker's code number if applicable.

8. COMPONENTS (ACTIVE)

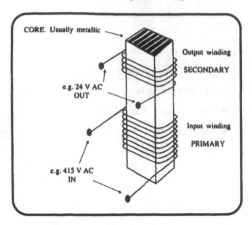

8.3. Transformers

A transformer is a device consisting of coils of wire wound on a metal core or frame and is used to convert AC voltages to higher or lower values.

For example, they are used to change the 415 V input voltage to 24 V for use by the contactor coils in the controller.

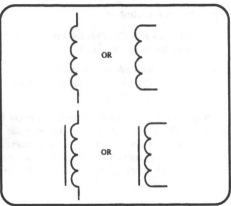

8.3.1. Symbols

The symbols used are closely allied to all components which use coils, i.e. *wound components* which include chokes, solenoids, relays, contactors and motors.

- General coil symbols.

- General symbols where the line indicates that the coil is wound on to a 'core'. Also used for *chokes or solenoids*.

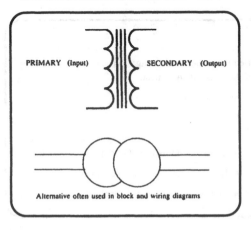

A basic isolating transformer has at least two separate windings – the primary which is the input and the secondary which is the output.

- This symbol shows a *double wound transformer*. These transformers are used to provide electrical isolation and also to provide the different voltage levels needed in the system.

- Double wound transformers are by far the most common in control systems.

8. COMPONENTS (ACTIVE)

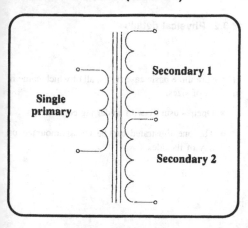

Secondary 1

Single primary

Secondary 2

- There can be a number of secondaries each one providing a different output voltage.

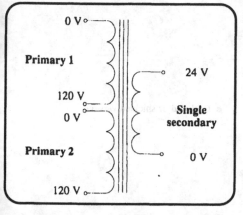

0 V

Primary 1

120 V

0 V

Primary 2

120 V

24 V

Single secondary

0 V

- There may also be more than one primary to allow the transformer to work at different input voltages or supplies.

- This shows two primaries, each working at 120 V.

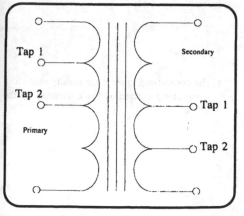

Tap 1

Tap 2

Primary

Secondary

Tap 1

Tap 2

- The windings may also have intermediate connections called *voltage taps* which can allow for variation of input voltages or fine adjustment of output voltages.

8. COMPONENTS (ACTIVE)

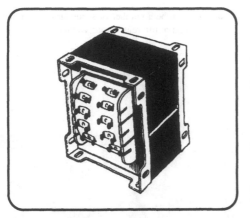

8.3.2. Physical details

There are three basic case styles, all of which come in a range of sizes.

- Open – usually with solder tag connections.
- The one illustrated can be chassis-mounted on any of its sides.

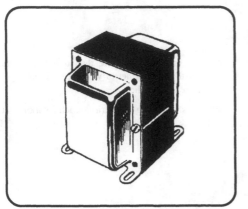

- Enclosed or shielded version.

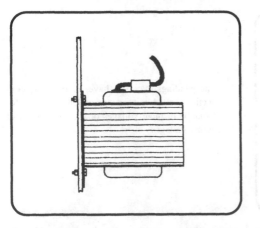

- The connections to these are usually made to a screw clamp connection block mounted on the side or top.

8. COMPONENTS (ACTIVE)

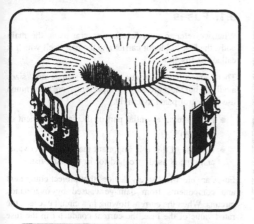

- Toroidal – connections may be to solder tags or flying leads.
- The leads will be identified by colour. A diagram to show which lead is which will be fixed to the transformer or supplied separately.

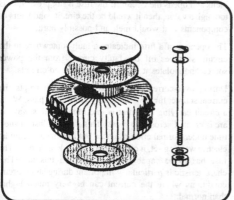

- The toroidal type is mounted using a long bolt and 'washers' or plates. Resilient pads should be used between the transformer and mountings.
- Do not overtighten the fixing screw since this could cause internal damage to the windings.
- There must be no conductive path between the central mounting bolt and the chassis, *around the outside of the toroid*. This would cause a short circuit and burn out the transformer.

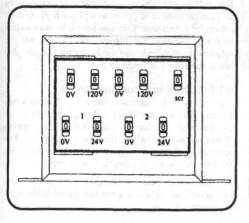

8.3.3. Markings

- The terminals may be marked with numbers or letters, in which case you need to have the information supplied by the maker unless your working drawing uses the same idents.

More common are voltage markings.

- **0 V** denotes the start of a winding, the higher voltage towards the end.
- **'scr'** is an interference screen which should be connected to the chassis.
- Never deviate from the wiring drawing since in most cases the terminals are not interchangeable.

91

9. COMPONENTS (PASSIVE)

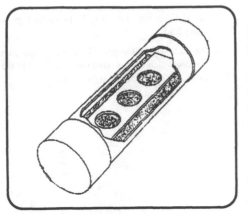

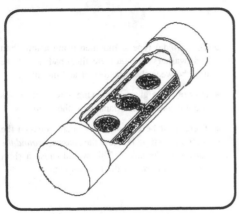

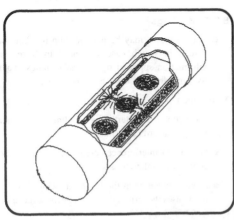

9.1. Fuses

What we refer to as a fuse has several parts, the main body, the fuse holder or carrier and the fuse itself which is called the fuselink.

There are a wide variety of types, shapes and sizes available but there are only a couple which are in common use in control panel assembly.

- Fuses are an essential part of the safety element of the equipment.

- Because of this it is important that the correct value and type is used as called for in the parts list.

Fuses are electrical safety devices that protect equipment and components from damage caused by overloaded circuits. When the current flowing in a circuit exceeds the rated value of the fuse, the current conductor in the fuse melts and opens the circuit. If the fuse is not present, or is too high a value, then it would be the circuit conductors or components that would melt and possibly burn.

The opening of a fuse indicates a fault somewhere in the circuit, switches off the faulty circuit from the power source and isolates it from other, unaffected circuits.

During over-current conditions the fuse interrupts the current source, limiting the energy allowed to pass. When a circuit carrying a current is interrupted in this way, an arc is created across the break. This arc only lasts a short time under normal circumstances but like the arc from an electric welding set, it can generate considerable heat. The fuse has to be capable of withstanding this arc. This characteristic is particularly important during short circuit conditions where the current can be very much higher than normal.

Fuse holders – or carriers – also have to be made so that they can carry the rated current as well as a high overload current for a short time. They also have to be made so that they can withstand the highest voltage they will be subjected to. Standards also dictate the type of fuse that has to be used for different circuits.

9.1.1. Soldering recommendations

Although not very common, some smaller fuses are soldered onto a circuit board and, since most fuses are constructed incorporating soldered connections, caution should be used when installing them into place. The application of excessive heat can reflow the solder within the fuse and change its rating and characteristics.

Fuses are a heat sensitive component similar to semi-conductors and the use of a heatsink during soldering is recommended.

9. COMPONENTS (PASSIVE)

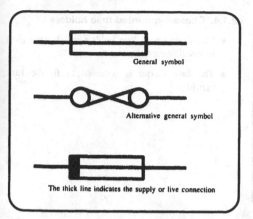

General symbol

Alternative general symbol

The thick line indicates the supply or live connection

9.1.2. Symbols

There are three symbols in common use.

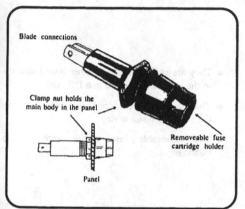

Blade connections

Clamp nut holds the main body in the panel

Removeable fuse cartridge holder

Panel

9.1.3. Panel-mounted fuses and holders

- They may be panel-mounted. This one has blade terminations for crimped spade connectors. Others have solder tags. There are various other similar body styles.

- Be careful when tightening the clamping nut. Overtightening will break the plastic body.

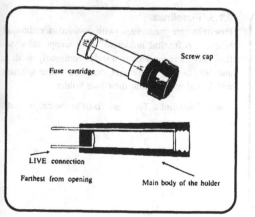

Screw cap

Fuse cartridge

LIVE connection
Farthest from opening

Main body of the holder

- When used for mains supplies, the *live* must be connected to the terminal which is in turn connected to the *inside* fuse contact, i.e. the one you can't touch with your finger! It is good practice to connect the supply side of any voltage source to this terminal.

- Fuse cartridges are usually 20 mm long.

9. COMPONENTS (PASSIVE)

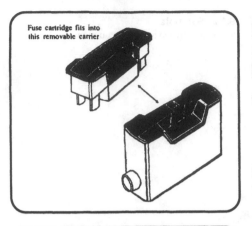

Fuse cartridge fits into this removable carrier

9.1.4. Chassis-mounted fuse holders

- Chassis-mounted fuse holders which have plug in fuselink carriers.

- The fuse carrier is removed to fit the fuse cartridge.

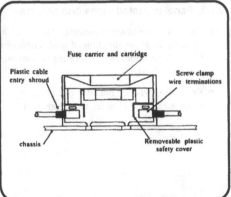

Fuse carrier and cartridge

Plastic cable entry shroud

Screw clamp wire terminations

chassis

Removeable plastic safety cover

- They are surface-mounted either bolted directly to the chassis or clipped to a DIN rail.

- These generally have screw clamp wire terminations for the panel wiring.

- The removable fuse carrier accepts fuse cartridges.

9.1.5. Fuselinks

Fuselinks are cartridges with welded termination brackets. A fuselink holder will only accept one style. Basically there are only two styles commonly used, A and NS, but be aware that there are some specials which will only fit into their own holder.

- 'A' fuselinks. These are fixed to the carrier with screws.

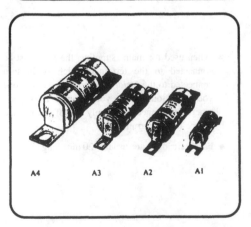

A4   A3   A2   A1

94

9. COMPONENTS (PASSIVE)

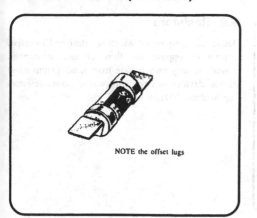

NOTE the offset lugs

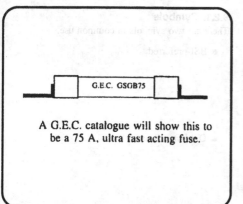

G.E.C. GSGB75

A G.E.C. catalogue will show this to be a 75 A, ultra fast acting fuse.

- 'NS' fuselinks which plug into slots in the contacts in the fuse carrier.

- The value of the fuse is given in amperes – abbreviated to amps or A.

- Fuselinks are available in a range of ampere values as well as a number of distinct types.

- They may be anti-surge (T), fast acting (F), High Breaking Capacity (HBC) or special semi-conductor types.

- Other features such as indicating when blown or special materials may also be called for.

- These attributes will only be indicated in the maker's code number which will also appear in the parts list.

- European standard fuses are now being used. The 'D' 'NH' and 'NEOZED' are the most popular.

9. COMPONENTS (PASSIVE)

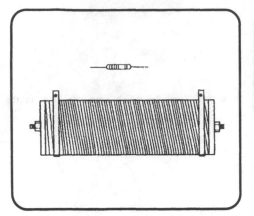

9.2. Resistors

These are components which are designed to resist, control or oppose the flow of electric current. Physically they vary in size from small (5 mm long) carbon devices to large wire-wound power resistors (up to about 300 mm long).

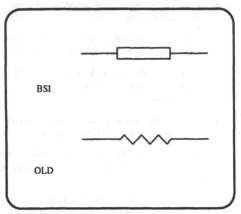

BSI

OLD

9.2.1. Symbols
There are two symbols in common use.

- BSI-preferred.

- Old but still used.

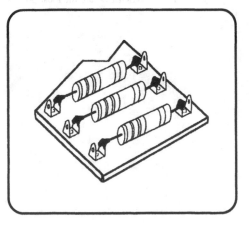

9.2.2. Fixed resistors

- Small wire-ended resistors are soldered to a printed circuit board or a tag strip to make a sub-assembly.

9. COMPONENTS (PASSIVE)

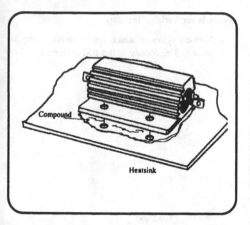

More common in control panels are wire-wound power resistors.

- This one is bolted flat to the chassis or more often a heatsink.
- To aid the transfer of heat from resistor to heatsink, a *heatsink compound* is used.
- The wires are soldered to the eyelets at either end.

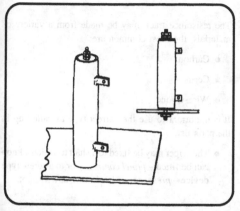

- This style is bolted to the chassis by a long bolt or stud through the middle.
- The connections are to the tags near each end of the body.
- Avoid overtightening which may cause damage.

Note that all resistors heat up in service and other parts, especially cables, should not be placed too close to them.

9.2.3. Variable resistors

These are mechanical devices where the resistance between a pair of terminals can be varied by moving a slider or wiper over a resistance track.

They are often called pots which is short for potentiometer. There are three terminals, one at either end of the resistance track and the other to the wiper.

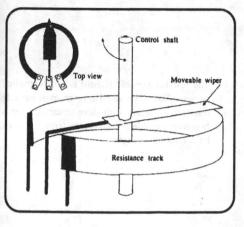

- This depicts a pot with a circular resistance track.

Some pots do not have a control shaft but are adjusted by the provision of a screwdriver slot. These are called trimpots.

9. COMPONENTS (PASSIVE)

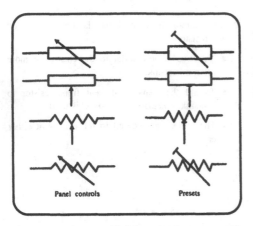

Panel controls   Presets

Symbols for variable resistors:

- Various symbols which are in common use are shown. The oblong is the BSI-preferred.

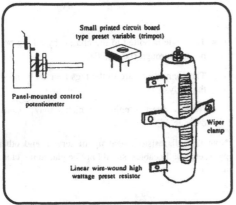

Small printed circuit board type preset variable (trimpot)

Panel-mounted control potentiometer

Linear wire-wound high wattage preset resistor

Wiper clamp

The resistance track may be made from a variety of materials, the most common are:

- Carbon.
- Cermet.
- Wire-wound.

It is important to use the correct type as called up in the parts list.

- The wiper may be fixed to a shaft to which a knob can be fitted – *panel controls* – or to a screw type device – *preset controls* – known as a trimpot.

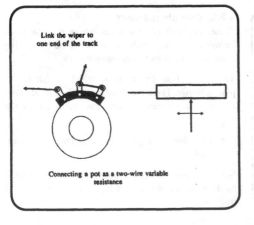

Link the wiper to one end of the track

Connecting a pot as a two-wire variable resistance

- The original variable resistor is a two-terminal device called a *rheostat*. However, most variable resistors are made with three terminals. For a two-wire variable resistance, the terminals must be connected as shown.

98

9. COMPONENTS (PASSIVE)

9.2.4. Resistor colour codes

4-band code

Colour	1st digit	2nd digit	Multiplier	Tolerance
Black	0	0	X 1	Not used
Brown	1	1	X 10	± 1%
Red	2	2	X 100	± 2%
Orange	3	3	X 1000	Not used
Yellow	4	4	X 10,000	Not used
Green	5	5	X 100,000	± 0.5%
Blue	6	6	X 1,000,000	± 0.25%
Violet	7	7	X 10,000,000	± 0.1%
Grey	8	8	Not used	Not used
White	9	9	Not used	Not used
Gold	Not used	Not used	X 0.1 or + 10	± 5%
Silver	Not used	Not used	X 0.01 or + 100	± 10%

Example 1

A = 1st digit is RED so is 2

B = 2nd digit is VIOLET so is 7

C = Multiplier is BROWN so is X 10

D = Tolerance is RED so is ± 10%

27 X 10 = 270.

This is a 270 ohm ± 10% resistor

Example 2

A = 1st digit is BLUE so is 6

B = 2nd digit is GREY so is 8

C = Multiplier is ORANGE so is X 1000

D = Tolerance is GOLD so is ± 10%

68 X 1000 = 68,000.

This is a 68,000 ohm ± 5% resistor

9. COMPONENTS (PASSIVE)

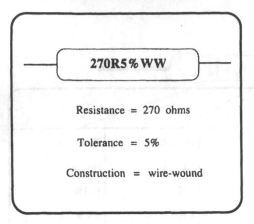

270R5%WW

Resistance = 270 ohms

Tolerance = 5%

Construction = wire-wound

9.2.5. Resistor value markings

The important parameters describing a resistor are:

- Resistance, measured in ohms, symbol Ω.
- Power measured in watts, symbol W.
- Construction or material.

Note 1. $1000 \text{ ohms} = 1000 \, \Omega = 1 \, k\Omega$

Note 2. Sometimes ohms (Ω) is written as R (see Section 9.2.7)

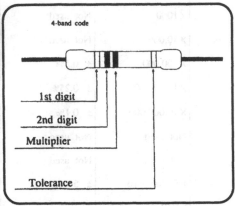

The resistor will be coded using the colour code shown on the previous page.

- This is marked on the resistor using four coloured bands.
- There is a wider gap between the first three bands and the last one.
- The first three denote the resistance.
- The fourth denotes a tolerance, i.e. how close the resistor may be to the marked value.
- This is a + or − figure.

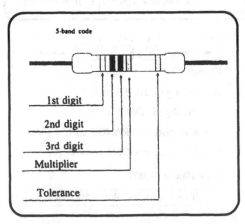

A variation to this adds a fifth band to the overall marking.

- Now four bands denote the resistance value. The last is still the tolerance.
- The fourth band is a third digit with the colours denoting the same value as the first two digits.

This allows more accurate values to be coded.

9. COMPONENTS (PASSIVE)

9.2.6. Temperature coefficient of resistance

A further variation in markings is to add yet another band on to the end to indicate the resistor's temperature coefficient, i.e. how much the resistance value changes with temperature.

All resistors change value as the temperature changes. Some types are more affected than others. When it is important that the effects are minimised, resistors with a small coefficient are specified by the additional colour band.

The first five bands are identical to the previous example which give the resistance and tolerance: a sixth band is added for the temperature coefficient.

The sixth band can be:	
Brown	200 ppm/°C
Red	100 ppm/°C
Orange	50 ppm/°C
Yellow	25 ppm/°C
Blue	10 ppm/°C
Violet	5 ppm/°C
White	1 ppm/°C

The ppm/°C stands for parts per million per degree centigrade. A 1 million ohm resistor with a temperature coefficient of 100 ppm would change by 100 ohms for every 1°C temperature change. The lower the figure the better the resistor's performance.

The decoding of these colour code bands is relatively easy. The main problem you will have will be making sure that you are reading the code the right way round.

Other problems come from the base colour of the resistor masking the code colour and distinguishing between orange, brown and red – colours are not very standard between manufacturers.

9.2.7. Alphanumeric resistor code

The colour code is not used in circuit drawings or parts lists. Power resistors, precision resistors and variable resistors may have their value *written* on.

The way in which the resistance is written is still in the form of a code. With this method – defined in BS1852 – the *multiplier* is given a *letter*.

● **R** is for the basic value in ohms where there is *no multiplier*, i.e. unity or '*times one*'.

● **K** – standing for *kilo*, and meaning '*times one thousand*'.

● **M** stands for *mega* and meaning '*times one million*'.

● **G** stands for *giga* and meaning '*times a thousand million*'.

● **T** stands for *tera* meaning '*times a million million*'.

47,000 Ω is written **47K**
237,000 Ω as **237K**
100 Ω as **100R**
1,000,000 Ω as **1M**

The *position* of the multiplier letter is used to denote the *position* of the *decimal point* in the resistance.

If the multiplier is at the end – as in **1R, 1K, 1M** then a **0** can be added after the multiplier – **1R0, 1K0, 1M0**.

The word ohms and its symbol are usually left off.

100 Ω would be marked **100R0**
2700 Ω (2.7 KΩ) as **2K7**
2.7 Ω as **2R7**

9. COMPONENTS (PASSIVE)

The *tolerance* is also given a letter:

F – 1%;

G – 2%;

J – 5%;

K – 10%;

M – 20%.

27K, 5% is written as **27KJ**

2R7, 10% as **2R7K**

237K, 1% as **237KF**

6M8, 20% as **6M8M**

9.2.8. Preferred values

An important fact is that not every value of resistance is made. Instead, a limited number of values are made. These are called preferred values and the number depends on the tolerance of the series.

By combining resistors any required value can be derived. In each tolerance band there are a set of *nominal values* and their multiples. The nominal values are such that the tolerance ranges will overlap the value above or below.

The 10% range is called the *E12* series since only 12 numbers (and their multiples) are required to provide a complete range of preferred resistance values:

1.0, 1.2, 1.5, 1.8, 2.2, 2.7, 3.3, 3.9, 4.7, 5.6, 6.8, 8.2.

By 'multiples' it simply means that resistors are made in sets of the above values multiplied by 0.1, 1, 10, 100, 1000 and so on.

For example, if you take the number 4.7 then, using the above multipliers, you can obtain resistor values of 0.47, 4.7, 47, 470, 4700, 47,000, 470,000, 4,700,000 ohms.

The 5% tolerance series is called E24 and there are 24 preferred values:

1.0, 1.1, 1.2, 1.3, 1.5, 1.6, 1.8, 2.0, 2.2, 2.4, 2.7, 3.0, 3.3, 3.6, 3.9, 4.3, 4.7, 5.1, 5.6, 6.2, 6.8, 7.5, 8.2, 9.1.

Using the same multiples as before you can see that a similar range of preferred values are obtained but with twice the choice of resistance.

The other popular series are the E48 series, 2% tolerance range with 48 nominal values and the E96 series, 1% tolerance range with 96 nominal values and multiples.

The use of a limited number of preferred values helps in colour code identification through familiarisation.

9.2.9. Variable resistor markings

The variable resistors may be marked with their resistance value in a similar way to power resistors to show the resistance and its tolerance.

However, there is another factor added. The resistance track can be made so that the resistance variation is *linear* or *logarithmic*.

- Linears are marked **linear, lin** or **ln**.

- Logarithmic are marked **log** or **lg**.

A 10,000 ohm, 10% pot where the resistance varied logarithmically would be marked:

10KK log.

The other parts of the specification are the power rating in watts and the track material. So the full specification for a 10,000 ohm pot with a carbon resistance track could be:

10K, 10%, log, 0.25 W, carbon.

Most preset pots are linear types.

9. COMPONENTS (PASSIVE)

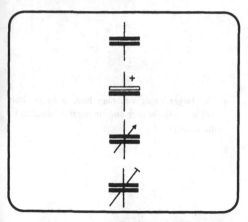

9.3. Capacitors

9.3.1. Symbols

This basic symbol for a capacitor or condenser is modified to show polarisation or variability when applicable:

- Polarised.

- Variable.

- Preset variable.

9.3.2. Physical details

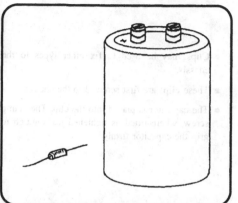

- They come in a wide variety of case styles and may also vary in size from the small electronic types of about 5 mm long to large components which resemble a can of beans!

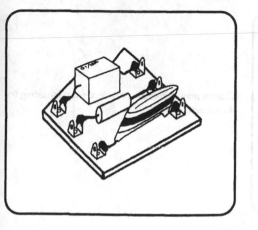

- Small capacitors are normally mounted to a tag strip as a sub-assembly. Three versions are shown.

103

9. COMPONENTS (PASSIVE)

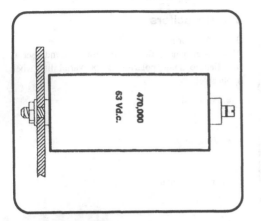

- The larger capacitors may have a single bolt welded to the bottom and are mounted directly to the chassis.

- Clips may be used to fix other types to the chassis.
- These clips are first screwed to the chassis.
- The capacitor is placed into the clip. The clamp screw, where fitted, is tightened just enough to grip the capacitor firmly.

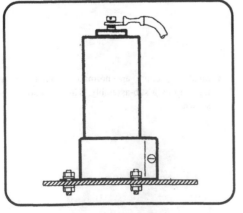

Clamps may be fitted to provide vertical mounting for the capacitor.

9. COMPONENTS (PASSIVE)

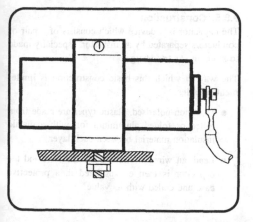

Or clips may be secured to give horizontal mounting.

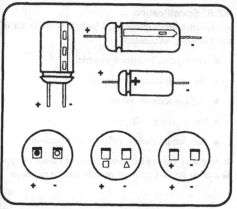

9.3.3. Polarity

From the wiring point of view there are two types and there is an important difference between them – *polarised and non-polarised*.

- A polarised capacitor is used in DC circuits and *must* be connected the correct way round.

- Polarised capacitors are marked in a variety of ways to indicate the polarity.

Non-polarised capacitors may be connected into the wiring either way round – it does not matter which terminal is connected to which wire.

9.3.4. Connections

Connections on larger capacitors are of three types:

- Screw clamp.

- Blade – for a crimp receptacle.

- Solder tag.

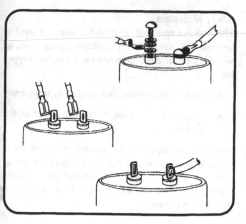

9. COMPONENTS (PASSIVE)

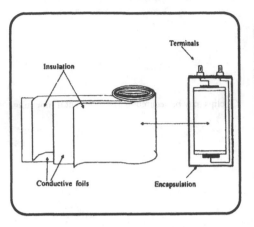

9.3.5. Construction

The capacitor is a device which consists of a pair of conductors separated by an insulator, especially made to store an electric charge.

The way in which this basic construction is implemented varies.

- Most non-polarised, plastic types are made from rolled or folded aluminium foil with a plastic insulation material between each layer.

- Lead-out wires are attached to the foils and the capacitor is then encapsulated in a protective case and coded with its value.

9.3.6. Specification

A capacitor specification will include some or all of the following parameters:

- The type of insulation material.

- The encapsulation material.

- The capacitance value.

- The working voltage.

- Any temperature coefficient of capacitance.

It may also include the construction or mounting type and other special information referring to its usage.

9.3.7. Materials

The material specified refers to the insulator used to separate the conductors. The encapsulating material will often be different. Common insulating layer types are:

- Plastic – polystyrene, polyester, polypropylene.

- Non-plastic – mica, ceramic, electrolytic, tantalum.

In *electrolytic* capacitors, the insulation layer is a thin oxide formed from an electrolytic liquid (or paste) when a voltage is applied. These and the *tantalum* capacitor have the characteristic of a high capacitance in a relatively small size.

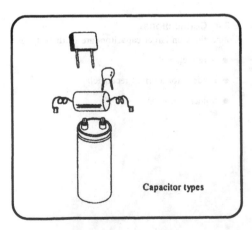

Capacitor types

9. COMPONENTS (PASSIVE)

They usually work properly at a specific voltage and, unlike the others, are polarised, meaning they must be fitted to a DC circuit the correct way round.

The insulation material and the encapsulation are chosen to suit a particular circuit application and the type must not be changed.

9.3.8. Capacitance values

The next part of the specification is the capacitance value itself. Before that, we need to look at the unit of capacitance and then see how it is marked on the capacitor.

The unit of capacitance is the farad.

The farad is a large unit when compared to the values in common use today and it is more likely that you will see values marked in smaller divisions or submultiples of a farad.

These are:

- microfarad – 1 microfarad is a millionth of a farad.

- nanofarad – 1 nanofarad is a thousandth of a microfarad.

- picofarad – picofarad is a thousandth of a nanofarad.

These are abbreviated to:

- 1 microfarad = 10^{-6} farads, shortened to μF.

- 1 nanofarad = 10^{-3} microfarads, or 10^{-9} farads, shortened to nF.

- 1 picofarad = 10^{-3} nanofarads, or 10^{-12} farads, shortened to pF.

Parts lists often use lower case 'u' or MFD for microfarad.

To convert capacitor values:

- pF to nF – divide by 1000. For example: $1000\,pF = 1\,nF$

- pF to μF – divide by 1,000,000. For example: $10,000\,pF = 0.01\,\mu F$

- nF to μF – divide by 1000. For example: $47\,nF = 0.047\,\mu F$

- μF to nF – multiply by 1000. For example: $0.022\,F = 22\,nF$

- nF to pF – multiply by 1000. For example: $22\,nF = 22,000\,pF$

10. SWITCHES AND LAMPS

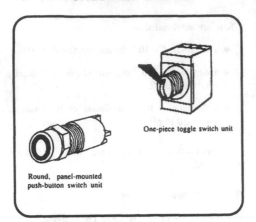

One-piece toggle switch unit

Round, panel-mounted
push-button switch unit

10.1. Switches

A switch consists of a set of contacts manually operated by some form of actuator.

The actuator and contacts may be contained in a single moulded unit or more likely as a modular unit comprising a selection of actuators and contact sets.

10.1.1. Moulded one-piece

- These are generally for low current use and are more likely to be found in the low voltage control system.

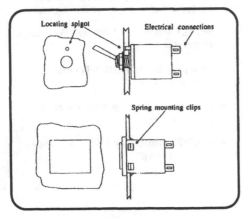

Locating spigot

Electrical connections

Spring mounting clips

Panel-mounted one-piece units are fixed to the panel using:

- either a central nut and lock washer – note the locating spigot, or
- clipped into a square hole.

The wires are generally connected using crimped spades although they can be soldered.

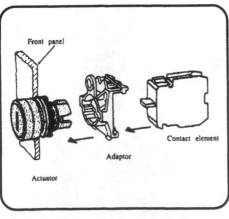

Front panel

Contact element

Adaptor

Actuator

10.1.2. Modular

These are built up using a choice of parts fitted to a panel-mounted body. The most popular size fits a 20.5 mm panel hole. Other sizes are 16 mm and 30.5 mm. While the actual detail of assembly varies between manufacturers, they are all similar to the following representative units.

There are three main parts:

- The actuator.
- Mounting adaptor.
- Contact elements.

108

10. SWITCHES AND LAMPS

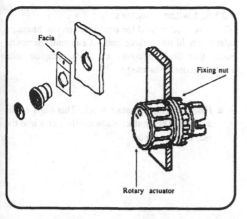

Facia

Fixing nut

Rotary actuator

10.1.3. Switch actuators

This is the part which will operate the switch contacts. There are several variations including some with lamp indicators. The actuator is fixed to the panel through a hole with a large fixing nut behind the panel. A lettered facia can be fitted between the flange on the actuator body and the panel.

- Rotary switch.

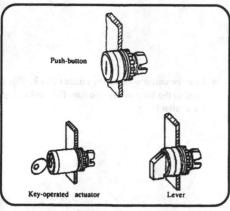

Push-button

Key-operated actuator

Lever

- Push-button switch.
- Key-operated switch.
- Lever switch.

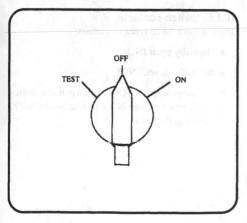

OFF

TEST

ON

10.1.4. Switch actions

- **Momentary** – where the contacts are operated only while the actuator is operated. Sometimes referred to as *spring return*.

- **Latching** – sometimes called *on-off* or, with a button actuator, *push on/push off*, where the contacts lock in one position when the button is pressed then released and only change back when the button is pressed a second time. *Stay-put* is yet another name.

Rotary actuators can provide more than two positions and may be used to provide a selector-type switch.

10. SWITCHES AND LAMPS

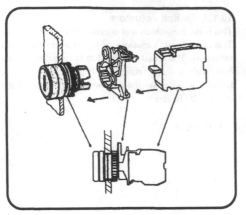

10.1.5. Switch adaptors

These are used to hold the contact elements. Standard adaptors hold up to three contact elements alongside each other. Some adaptors are made complete with contacts (contact blocks).

- Front-mounting contact block. This clips to the actuator. The contact elements then clip into the rear of the adaptor.

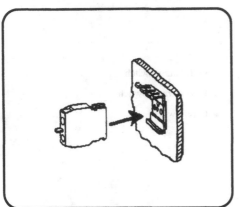

- Rear- or surface-mounting contact block. This is fixed to the base of the housing. DIN rail fittings are available.

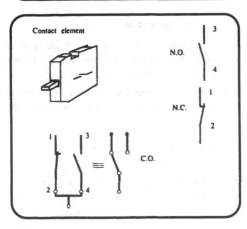

10.1.6. Switch contacts

There are two basic types of contact:

- Normally open (NO).

- Normally closed (NC).

- A changeover set (CO), can be made from a combination of one NO and one NC by wiring them together as shown.

10. SWITCHES AND LAMPS

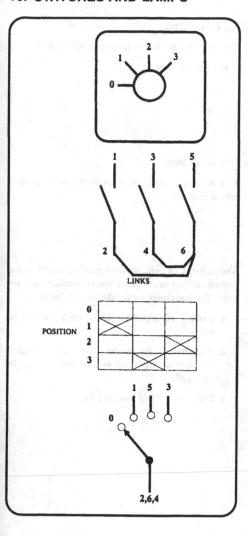

LINKS

POSITION

10.1.7. Rotary switch diagrams

The contact diagrams for rotary switches are often accompanied by an operational grid showing which contact operates in each position.

- Front panel view of a 4-position rotary switch with an 'off' position.

- This is sometimes referred to as a 0 position, 1 pole, 3 step switch.

- This is the circuit diagram showing the individual switch elements, in this case all NO.

The grid. The large cross in a contact square indicates that it is operated. Absence of the cross means it is not operated. This example shows that:

- No contacts are operated in the 0 position.

- Contact 1,2 operates in the 1 position.

- Contact 5,6 operates in the 2 position.

- Contact 3,4 operates in the 3 position.

- In all positions only one contact is operated.

The bottom drawing shows an alternative way of representing the same rotary switch.

Note that in a circuit diagram the individual switch contacts may be drawn in different parts of the drawing and will then be identified by a switch reference number as well as the contact numbers.

10. SWITCHES AND LAMPS

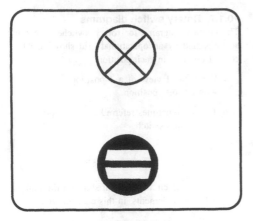

10.2. Lamps

Two symbols are shown recommended by BSI.

- Indicator lamp.

- Signal lamp.

It is not important from an assembly point of view which is which.

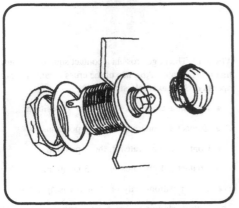

The majority of indicators are panel-mounted. These consist of two main parts, the lampholder and the bulb. The **lampholder** can take several forms:

- One-piece holder which fits through a hole in the panel.

- A nut holds it tight to the panel. Take care not to overtighten this otherwise the holder may be damaged.

- The bulb is fitted from the front.

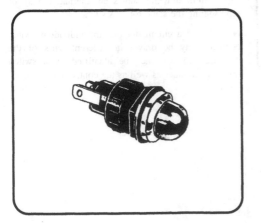

- This type is very similar and fits in the same way but the bulb is fitted from the rear.

10. SWITCHES AND LAMPS

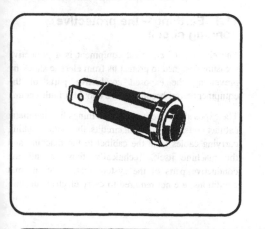

- The neon type works at high voltages – more than 100 V – and is often not removable from the holder.

- Connections to the above three holders may be crimped blade or soldered joints.

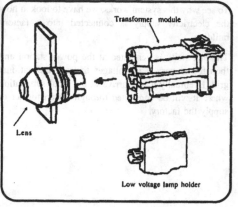

Transformer module

Lens

Low voltage lamp holder

- Another popular type has several parts to assemble in an arrangement similar to switches and there are some which contain a transformer so that they can work from the mains supply.

- This holder is the same size as a switch element, uses low voltage bulbs and clips to the rear of the lens holder.

- The terminations are screw clamp.

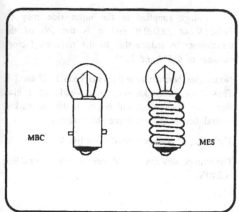

MBC

MES

- Filament bulbs usually operate on 12 V or 24 V supplies and may be a screw-in type (MES) or bayonet cap (MBC).

Note. The lens colour will be specified and should not be altered. (The colour signifies a particular condition to an operator of the finished equipment.) The bulbs will also be specified in terms of voltage and power.

11. EARTHING AND SCREENING

11.1. Earthing – the protective bonding circuit

The earthing of electrical equipment is a protective measure designed to protect us from electric shock by preventing the exposed conductive parts of the equipment from becoming live should a fault occur.

The exposed conductive parts are things like the metal cabinet housing the control circuits, the metal trunking carrying cables from the cabinet to the machine and the machine itself. Technically, they are all the conductive parts of the system that, under normal conditions, are not required to carry electric current.

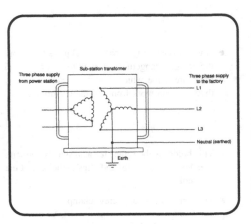

To see why this system works, we have to look at how the electricity supply is connected into a factory building.

The electricity is generated at the power station and then fed at very high voltages through the national grid. Eventually it will arrive at the local sub-station where it will be connected through a transformer to supply the factory.

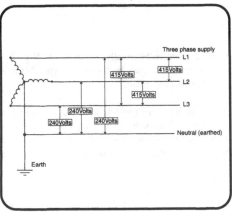

The voltage supplied to the input side may be 11,000 V or 33,000 V and it is the job of the transformer to reduce this to the required factory voltages of 415 V and 240 V.

Notice that there are three lines marked L1, L2 and L3. These three lines are known as phases and in effect give three 'live' connections with respect to the line marked neutral, hence the term 'three phase supply'.

The voltage between each of these live lines is 415 V.

The voltage between each of them and the neutral line is 240 V.

11. EARTHING AND SCREENING

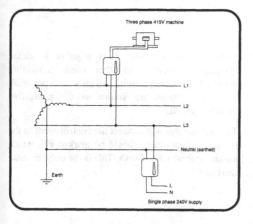

Three phase 415V machine

L1
L2
L3
Neutral (earthed)
Earth
L
N
Single phase 240V supply

The three phase supply at 415 V is used to power most electrical machinery in the factory since it is able to supply large currents.

The lower voltage between one phase and neutral is connected to the lights and 13 amp sockets as the 'single phase' supply just as we have at home, with a single live connection and a neutral connection.

Look back at the sub-station transformer diagram and see that the neutral point is also connected to the earth. This is usually accomplished by burying a large copper plate under the ground.

This means that as well as the live lines being at 240 V above the neutral line they are also 240 V above the earth. The earth in this case means not only the earth terminal in the supply box but the ground you stand on as well.

It also means that if you are in contact with the ground and also touch a live connection, you would receive a dangerous electric shock.

However, this earth connection is there to provide a safety function and to see how it does so, consider the following situation.

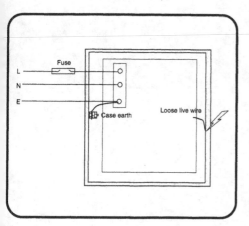

Fuse
L
N
E
Case earth
Loose live wire

This is the basic wiring diagram of a piece of mains-powered equipment which is contained in a metal cabinet.

Should a fault occur which causes the metal case for instance to be connected to the live side of the mains supply, a large fault current will flow into the earthing system. This current will cause the supply fuses to blow thereby disconnecting the electrical supply to the system. The case would therefore never become live.

Without the earth connection, the metal case would simply become live, possibly without causing any apparent fault in system operation. However, anyone touching the case could receive an electric shock.

11. EARTHING AND SCREENING

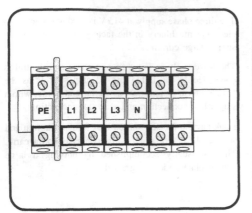

Connecting all of these parts together is called equipotential bonding. In other words connecting them all to the same potential – usually earth potential. These connections are known as the **protective bonding circuit**.

The terminal that will connect the control panel to the incoming supply earth should be marked **PE**, which stands for **Protective Earth**. This is the only terminal marked PE.

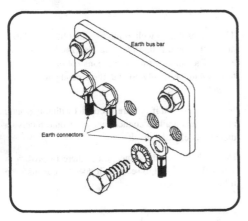

The protective earth terminal is then connected to the enclosure case, the chassis and to other equipment which has a metal case or chassis. This is usually done through an earth busbar.

11. EARTHING AND SCREENING

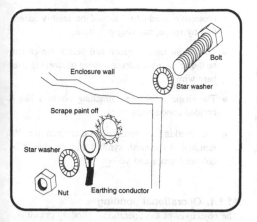

If there is no welded earth stud on the enclosure wall it will be necessary to connect the earth lead to a bolt. If necessary drill a suitable hole and scrape off any paint or other insulating coatings from both sides of the hole to give a good conductive path. Use washers and nuts as shown.

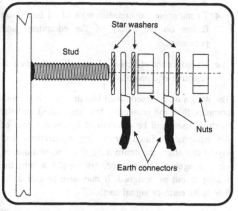

Where there is a stud and you want to connect two earth leads to it. Do not lay one lug on top of the other. This type of connection can work loose due to compression of the terminal eyelets. The correct way is to sandwich the first eyelet with star washers and a nut. After tightening the first nut, sandwich the second eyelet between it and a star washer and second nut.

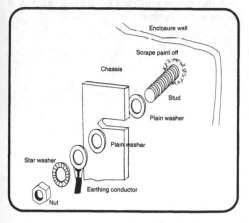

When you are mounting a chassis or equipment mounting bracket to an enclosure which has welded mounting studs, it is necessary to ensure a good connection to earth by adding a separate earth wire. Use the sequence of washers and nut as shown. If the chassis or bracket are coated then scrape it off to provide a good connection.

11. EARTHING AND SCREENING

Because of the safety aspects of the protective bonding circuit, there are a number of recommendations and requirements contained in both standards and regulations. While most concern the designer of the equipment, we should also be aware of some of them.

- All exposed conductive parts of the electrical equipment and the machine must be connected to the protective bonding circuit.

- Metal conduits and metal sheathing of cables should be connected to the protective bonding but should not be used as part of it. A separate earth wire should be connected and carried along the conduit.

- Where there is electrical equipment mounted on doors, lids or cover plates these should be connected to the protective bonding with an earth strap. The hinges or other fastenings are not reliable enough to ensure a good connection.

- No switching devices should be included in the protective bonding other than links that will have to be removed using a tool.

- It is not necessary to connect small parts such as screws, rivets and nameplates to parts mounted inside an enclosure such as the metal parts of components.

- Where a sub-assembly is connected using a plug and socket combination, the type used should ensure that the protective bonding circuit is the last to be disconnected. Usually this means that the earth pin is longer than the others.

- Protective conductors should be readily identified by shape, marking or colour.

- The colour used is green and yellow which can be the cable insulation or a piece of sleeving over bare wire.

- The shape would be something obvious like a braided conductor.

- The marking of terminals other than the PE terminal is the earth symbol or they should be coloured green and yellow.

11.1.1. Operational bonding

The objectives of the operational bonding circuit:

- To minimise the consequences of an insulation failure on the safety of the equipment and personnel.

- To reduce the effects of electrical interference on sensitive equipment.

The design of the operational bonding circuit is more complex than the protective bonding circuit and the methods used will be determined by the designer of the equipment. Since the wiring connections and layout can affect the conformity to the regulations on electromagnetic interference, we should adhere the wiring layout as designed. It may also be called the low-noise earth or signal earth.

Note that at some point these two earthing circuits will be connected together.

11. EARTHING AND SCREENING

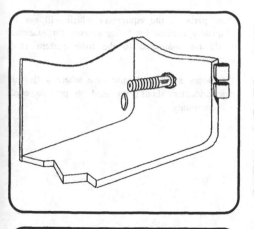

11.1.2. Cabinet earthing

- Metal enclosures will have an earth terminal stud welded to the inside.

- This must be connected to the *main supply earth terminal*.

- Sub-assemblies in metal enclosures will also have an earth terminal.

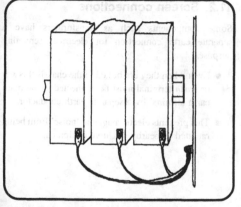

11.1.3. Earth continuity conductor

Where there are several sub-assemblies which require earthing in the enclosure each must be independently connected to the main earth terminal so that if one unit is removed the others remain earthed.

- The earth connection may be made using separate flexible leads – often braided.

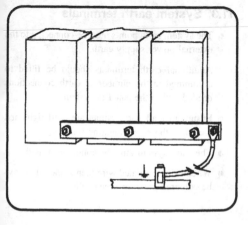

- Alternatively a solid copper busbar may be used.

11. EARTHING AND SCREENING

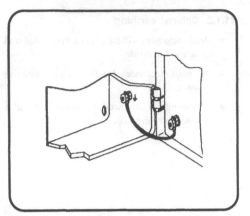

Other parts of the equipment which will not be adequately earthed by fixing screws, for example, should be connected to the main system earth terminal.

- Doors are a particular case where a flexible conductor should be used to provide earth continuity.

11.2. Screen connections

Some components such as transformers have a separate earth connection for electrical screening purposes.

- Even when they are bolted to the chassis this *scn* or *earth* terminal must be connected to the main earth terminal by a separate earth conductor.

- This prevents electromagnetic 'noise' from being radiated to nearby sensitive equipment.

11.3. System earth terminals

- The system earth points must be connected to the external power supply earth.

- Additional earth terminals should be fitted to accommodate the number of earth connections needed within the panel or system.

- Ensure that all earth points are well tightened including those on mounting rails.

- The earth system must be continuity tested.

- When using insulated wire then it should always be coloured green or green/yellow.

12. PLC WIRING

The PLC – programmable logic controller – is an industrialised computer designed specifically for industrial control systems.

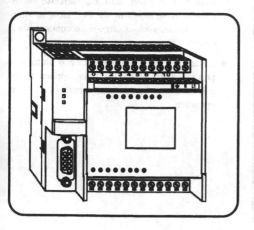

There are two main styles:

- A small so-called all-in-one type, designed to be mounted directly into the panel on a DIN rail. These are wired in a similar way to contactors.

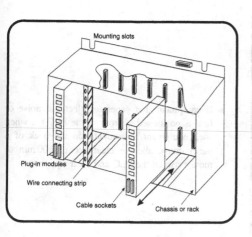

Mounting slots

Plug-in modules

Wire connecting strip

Cable sockets

Chassis or rack

- A modular type, which consists of a frame or rack and a number of separate, plug-in modules. The rack is first fixed on to the chassis – the smaller type fit on to a DIN rail, the larger type require bolts. The separate modules are then plugged into this rack.

Cable connections may be to screw terminals mounted on the rack or use multi-way connectors that plug into each module. The rack should be connected to the protective bonding circuit.

12. PLC WIRING

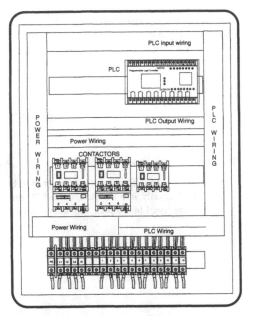

12.1. Installation

- They are generally robust but they contain electronic components and printed circuit boards so have to be fixed into a control cabinet away from heat, moisture, dust and corrosive atmospheres. Avoid mounting the PLC close to vibration sources, such as large-sized contactors and circuit breakers.

- Ensure that the mounting surface is uniform to prevent strain. Excessive force applied to the printed circuit boards could result in incorrect operation.

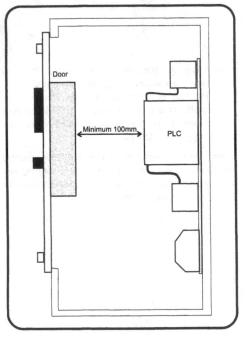

- If equipment that generates electrical noise or heat is positioned in front of the PLC (as when such equipment is mounted on the back of a panel door), allow a clearance of 100 mm or more between the PLC and such equipment.

12. PLC WIRING

12.2. Power supply wiring

When wiring AC supplies:

Live – L
Neutral – N
Earth – E or symbol

With DC supplies:

Positive cable to +
Negative cable to –

DC power supply connections must never be reversed.

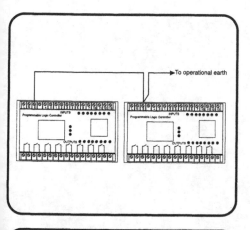

12.3. Earthing

- Use $2\,\text{mm}^2$ wire or larger.

- The PLC's earth should not be directly connected to the same terminal as power devices, but should go to the operational earthing circuit. The PLC will work without an earth connection but it may be subject to malfunction due to electrical interference.

- Where several PLCs or expansion units are used, link all the earth terminals together then connect to the operational earth terminal.

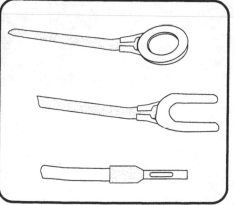

- Where the PLC connections are by screw terminals, use an insulated crimp eyelet or spade terminal on the cable end.

- Where the connections are to a terminal block type connecter, use an insulated ferrule on the cable end.

- Make sure the terminal screws are tightened correctly.

12. PLC WIRING

12.4. Wiring of inputs and outputs

- It is advisable to keep the input and output (I/O) lines separated from each other if possible.

- When the I/O signal wires cannot be separated from the power wiring, use shielded multi-core cable earthed at the PLC end.

- I/O signal wires should be kept at least 100 mm away from high-voltage and large-current, power circuit wiring.

INDEX

Printed in the United States
by Baker & Taylor Publisher Services